ENZYMATIC HYDROLYSIS OF CELLULOSE

ENZYMATIC HYDROLYSIS OF CELLULOSE

Theory and Applications

by

C.R. Wilke
B. Maiorella
A. Sciamanna
K. Tangnu
D. Wiley
H. Wong

Lawrence Berkeley Laboratory
and
Department of Chemical Engineering
University of California
Berkeley, California

NOYES DATA CORPORATION

Park Ridge, New Jersey, U.S.A.

1983

Library of Congress Catalog Card Number: 83-2294
ISBN: 0-8155-0945-6
ISSN: 0198-6880
Printed in the United States

Published in the United States of America by
Noyes Data Corporation
Mill Road, Park Ridge, New Jersey 07656

10 9 8 7 6 5 4 3 2 1

Library of Congress Cataloging in Publication Data
Main entry under title:

Enzymatic hydrolysis of cellulose.

(Chemical technology review ; no. 218)
Bibliography: p.
Includes index.
1. Cellulose--Chemistry. 2. Hydrolysis.
I. Wilke, Charles R., 1917- II. Series.
TS933.C4E59 1983 661'.802 83-2294
ISBN 0-8155-0945-6

Foreword

This book reviews the theory and applications of enzymatic hydrolysis of cellulosic biomass, a renewable raw material for chemicals and energy. The subject has been of great interest during recent energy crises; but it may possibly be of greater interest as genetic engineering techniques are applied to the conversion of cellulose to sugars.

In either case, this excellent and thorough study fills a gap in the available literature on the subject. The state of the art and potential industrial processes are detailed. A section is also included on high productivity fermentation systems for the production of ethanol.

The information in the book is from *Enzymatic Hydrolysis of Cellulose: Theory and Applications* (LBL Report 13669), prepared by C.R. Wilke, B. Maiorella, A. Sciamanna, K. Tangnu, D. Wiley, and H. Wong of Lawrence Berkeley Laboratory and the University of California Department of Chemical Engineering, for the U.S. Department of Energy, June 1980.

The table of contents is organized in such a way as to serve as a subject index and provides easy access to the information contained in the book.

Advanced composition and production methods developed by Noyes Data are employed to bring this durably bound book to you in a minimum of time. Special techniques are used to close the gap between "manuscript" and "completed book." In order to keep the price of the book to a reasonable level, it has been partially reproduced by photo-offset directly from the original report and the cost saving passed on to the reader. Due to this method of publishing, certain portions of the book may be less legible than desired.

Notice

Contents and Subject Index

Introduction

As a renewable raw material source for chemicals and energy, biomass has much promise. Grains and sugar, for example, can readily be converted to ethanol for use as a chemical intermediate or motor fuel. But longer range world needs will impose a limitation on the use of such crops for purposes other than food.

The cellulosic portion of biomass, however, represents an immense source of sugars which awaits only the development of the technology necessary for its economical utilization. Typically about 40% of the content of plant tissue is comprised of cellulose, a polymer of hexose sugars, and about 20% consists of hemicellulose, largely a polymer of pentose sugars. The monomer sugars can be produced by hydrolysis of the cellulosic fractions employing as catalysts either acids or enzymes. In light of an annual world production of biomass on the order of 10^{11} metric tons, the tremendous potential for a sugar based chemical and energy industry is apparent.

Enzymatic hydrolysis of cellulose is attractive because of its specificity and absence of the competitive degradation which normally accompanies acid hydrolysis. There has been a growing emphasis in research on enzymatic hydrolysis in recent years, stemming importantly from the pioneering work of Reese and Mandels at the U.S. Army Natick Laboratories. Over the last decade research in this field has been stimulated in the U.S. by major support from the National Science Foundation, the Department of Energy and its predecessor organizations, and most recently by the Solar Energy Research Institute. Research and development is actively underway in many other countries as well.

A large amount of research has been published on the theory of enzymatic hydrolysis and the various microbial, and other, sources of the enzymes. The present report endeavors to supplement this information by emphasizing insofar as possible the status of the technology and of potential industrial processes for production of sugars from cellulose. A substantial research effort on cellulose conversion has been underway in the authors' laboratories at the University of California at Berkeley over the past ten years. This report is based in part upon this background of experience, and experimental data from relatively recent studies are presented in certain sections to make the information as timely and useful as possible.

Because of current interest in production of ethanol a section is included which summarizes various methods for high productivity fermentation systems.

I. Theory of Enzymatic Hydrolysis

A. Sources of Cellulases

Microorganisms which can grow on cellulose include true bacteria, actinomycetes and higher fungi. Those which can utilize native cellulose rather than only soluble derivatives are termed truly cellulolytic.

Fungi which have been studied for their cellulolytic ability include *Irpex lacteus* (1), *Pyricularia oryzae* (2), *Penicillium funiculosum*, *Fusarium solani*, (3) *Chaetomium thermophile*, *Myrothecium verrucaria*, *Phanerochaete chrysosporium* (*Sporotrichum pulverulentum*)(4-6), *Trichoderma koningii* (7,8), and *T. reesei* (*T. viride*). Much of the recent work published on cellulose degradation has been based on studies of the high-activity enzymes produced by the last three organisms.

From the wild type of *T. reesei*, isolated by the U.S. Army Natick Research and Development Command and designated QM-6a mutants have been developed. Of particular interest are the strains QM-9414 (ATCC 26921) and strains produced at Rutgers: Rut-NG14 and Rut-C-30 (32). The last strain is both hyperproducing and catabolite repression resistant and has high xylanase activity. The geneology of these strains is shown in Figure I-1.Additional strains are being investigated for β-glucosidase resistant to end product inhibition.

B. Cellulolytic Enzymes

The first postulate concerning the nature of the mechanism of enzymatic hydrolysis was advanced by Reese, *et al.* (9). This was based on the fact that some organisms could hydrolyze native cellulose, while others degraded only soluble derivatives. A two-step process was envisioned. A component termed C_1 was thought to initiate hydrolysis by a preliminary activation or disaggregation of the cellulose chains. Subsequently, C_x, the second component, was responsible for the depolymerization to soluble cello-oligosaccharides.

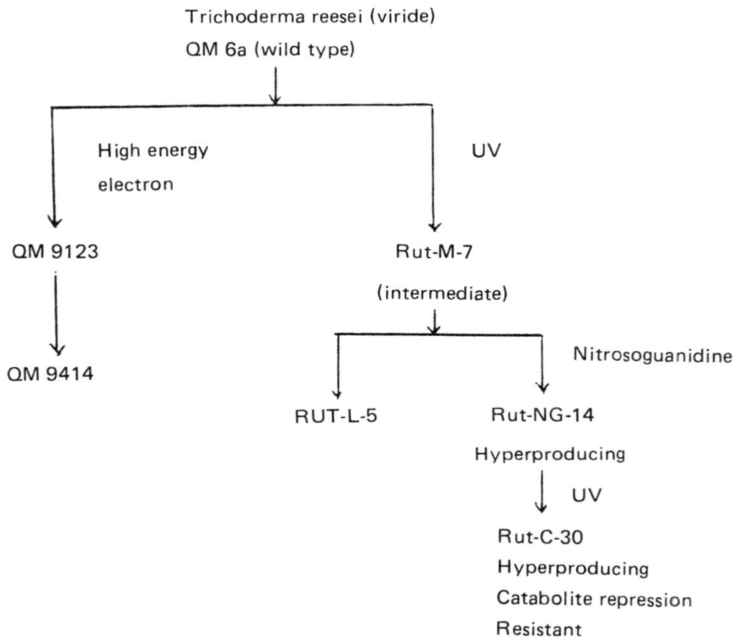

Geneology of High Yielding Cellulase Mutants

Figure I-1

Supposedly, truly cellulolytic organisms possessed the C_1 enzyme, which other cellulose degrading organisms lacked.

Later work has shown, in contrast, that C_1 is a β-1,4-glucan cellobiohydrolase (10,11), and thus acts on the chains formed by C_x action, contrary to previous conjecture. Fractionation resulting in purification of these enzymes have been performed by numerous investigators (12-17). Purified components retain both C_1 and C_x activities, although relative values of each activity differ.

Enzymes designated C_1 and C_x show a large synergistic effect, the exact nature of which has not been ascertained. After purification, activity towards native cellulose is decreased. Subsequent combination of the purified enzymes in their original proportions results in the recovery of the original high activity.

C. Composition of the Cellulase System

The cellulase system has been shown to exhibit three distinct types of cellulolytic activity which bear the descriptive names: β-1,4-glucan glucanohydrolase (EC 3.2.1.4) an endo-enzyme; and β-1,4-glucan cellobiohydrolase (EC 3.2.1.91), an exo-enzyme; and β-glucosidase (EC3.2.1.21). These more exact names for the endo- and exoglucanases should replace the trival names C_1 and C_x.

The first enzyme is a randomly-acting endoglucanase, i.e. it acts on the interior of the polymer to generate new chain ends. This activity is assayed by reactivity towards soluble cellulose as shown by increases in reducing sugars or by decreases in viscosity.

The second enzyme acts on the nonreducing ends of the polymer chain to release cellobiose. This activity is determined by incubation with crystalline cellulose. When highly purified, it shows only slight activity due to the synergism of the enzyme system.

β-glucosidases hydrolyze cellobiose, and less actively short-chain cell-oligosaccharides, and are necessary for removal of product which would otherwise inhibit the progress of the hydrolysis. Recent investigation has suggested that β-glucosidase activity as determined by reaction with p-nitrophenyl-β-glucoside must be distinguished from activity towards cellobiose. The mode of action of the separate activities of the combined system is shown in Figures I-2 and -3, respectively.

D. Fractionation of Cellulases

Cellulases have been fractionated into their constituent components by a variety of biochemical procedures including ultrafiltration, gel filtration, ion-exchange chromatography, adsorption chromatography and iso-electric focusing Endoglucanases and exoglucanases have been found to be glucoproteins. Isozymes have been reported for each component.

The cellulase of *T. koningii* has been fractionated by Wood and McCrae (12,18) into eight components including a single exoglucanase, five endoglucanases and two β-glucosidases. *P. chrysosporium* cellulase has been fractionated by Eriksson and co-workers (406), resulting in the isolation of an exoglucanase, five endo glucanases, two β-glucosidases and a cellobiose oxidase.

The separation schemes of Berghem and Pettersson (19-21) and of Shoemaker, Gum and Brown (22-24) for the fractionation of cellulase of *T. reesei* are shown in Figures I-4 and -5, respectively. From industrial preparations of cellulases, Berghem and Pettersson have reported isolation of two endoglucanases and a single exoglucanase and β-glucosidase, while Shoemaker, *et al.* have reported four endoglucanases, one exoglucanase and one β-glucosidase. The latter report a separation suitable for a large-scale purification procedure.

Physical properties of cellulase constituents prepared by various investigators are shown in Table I-1.

ENDOGLUCANASES:	Random action on amorphous cellulose	C_x G-G-G-G↓-G-G-G-G-G-G
EXOGLUCANASES:	Endwise action on crystalline and amorphous cellulose	C_1 G-G↓-G-G-G-G-G-G-G-G
β-GLUCOSIDASES:	Hydrolysis of cellobiose to glucose	β G↓G

Figure I-2. Separate Activities of the Cellulase Complex

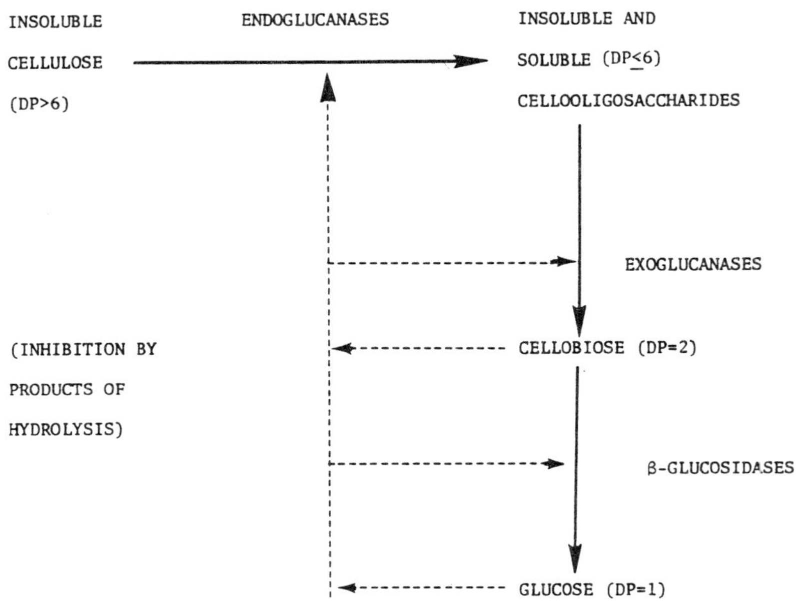

MODE OF ACTION OF CELLULASE

Figure I-3.

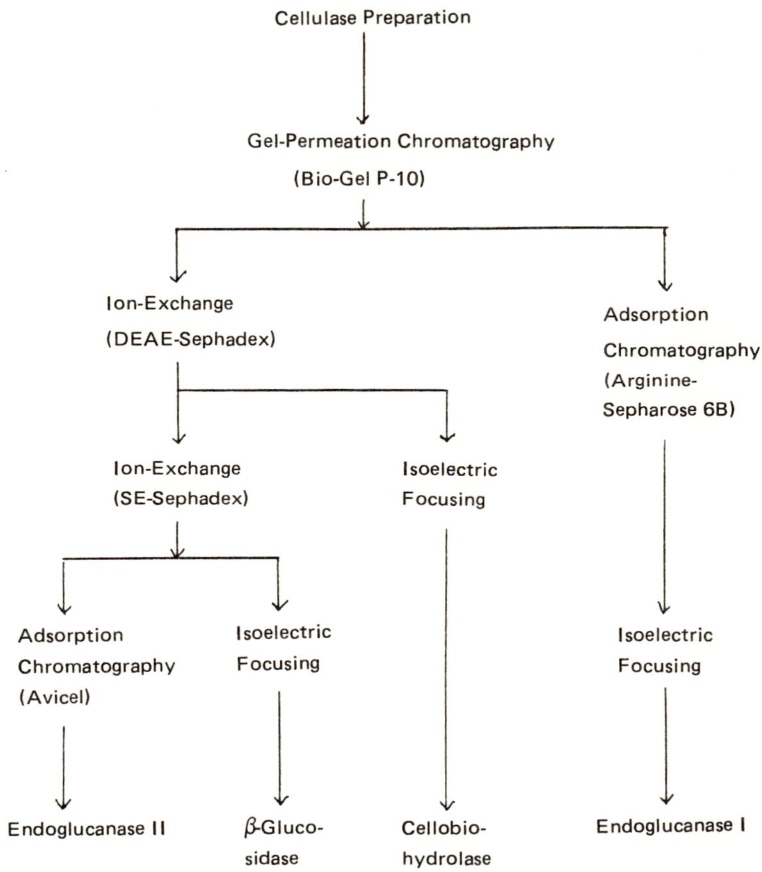

Cellulase Fractionation (Berghem, Pettersson)

Figure I-4.

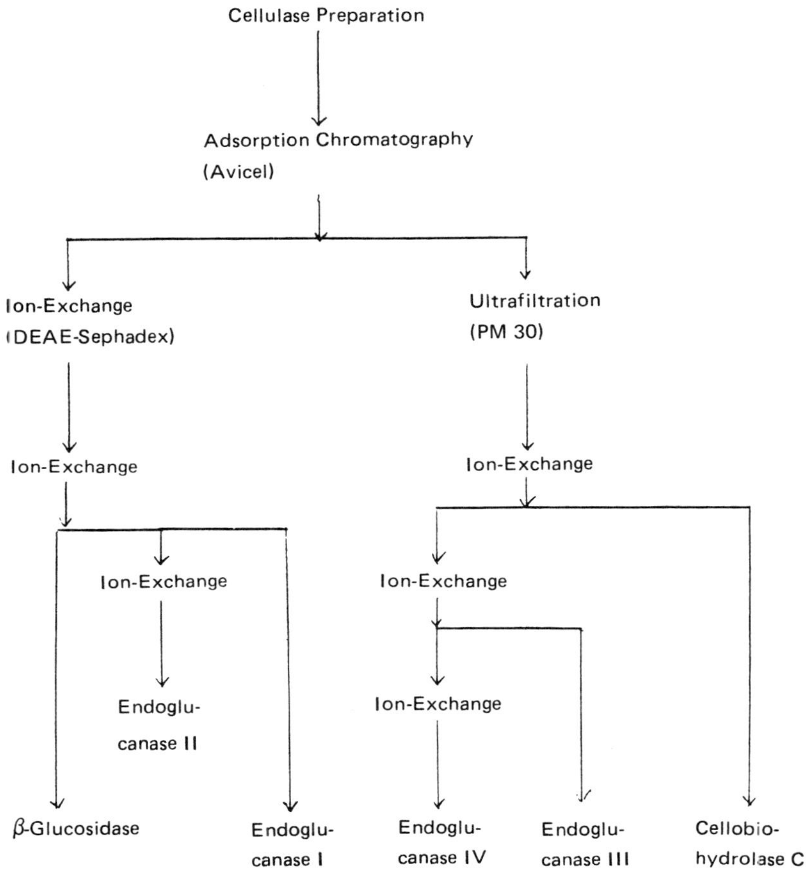

Cellulase Fractionation (Shoemaker, Gum, Brown)

Figure I-5 .

Table I-1

Properties of cellulase components of *Trichoderma reesei*

Component	Molecular Weight	Isoelectric Points	Reference
Exoglucanases	49,000		(9)
(EC 3.2.1.91)	61,000		(8)
	46,000	3.79	(16)
	48,000		(20)
Endoglucanases	76,000		(9)
(EC 3.2.1.4)	30,000; 43,000		(1)
	12,500; 50,000	4.60, 3.39	(18)
	37,200; 52,000;		(21)
	49,500		(21)
β-Glucosidase	47,000	5.74	(17)
(EC 3.2.1.21)	76,000; 76,000		(13)

E. Models of Enzymatic Hydrolysis

A model for the enzymatic hydrolysis of cellulosic materials must take into account the effects of the physical structure of the substrate, the nature of the cellulase complex, and the inhibitory effects of both substrates and products, including material present in the substrate other than cellulose, i.e. hemicellulose and lignin.

The major structural features that determine susceptibility to enzymatic degradation are crystallinity (25,25), and accessibility, which is defined by the surface area accessible to enzymatic attack. Thus pretreatment has a profound effect on hydrolysis.

The kinetic behavior of enzymatic degradation is complicated by the insoluble nature of the substrate, so that adsorption of enzyme and diffusion of intermediate products may be of importance. Using cane bagasse and cellulases of T. reesei mixed with xylanase of Aspergillis wentii, Ghose and Bisaria (27) have shown that endoglucanase adsorbs on cellulose preferentially to exoglucanase, with β-glucosidase not adsorbing to a significant extent.

Huang (28), assuming the fast adsorption of two enzyme components (C_1 and C_x) with subsequent product inhibition by both cellobiose and glucose, devised a kinetic model which was able to predict the rate of hydrolysis of amorphous cellulose (Solka Floc) up to about 70% conversion.

Okazaki and Moo-Young (29) developed a model of cellulose degradation to explain the synetgistic effect of cellulases, dependency of hydrolysis rate on the degree of polymerization of the substrate, and the effects of the substrate inhibition. The model included the effect of the three types of enzymes known to be actually present.

Howell and Stuck (30) presented a model neglecting substrate multiplicity and differences between effective cellulose concentration and average bulk concentration, and including the inhibitory effect of cellobiose. They showed that noncompetitive inhibition by cellobiose dominates the reaction kinetics. This model was then extended to higher conversions by postulating inactivation of the adsorbed enzyme-substrate complex (31).

Because of the complex nature of the multiple reactions, further investigation is necessary in developing a general model to describe the cellulose-cellulase system.

References

1. Kanda, T., et al., J. Biochem. 84, 1217-1226 (1978).

2. Hirayama, T., J. Biochem. 84, 27-37 (1978).

3. Suga, K., et al., Biotechnol. Bioeng. 17, 189-201 (1975).

4. Eriksson, K-E., Biotechnol. Bioeng. 20, 317-332 (1978).

5. Deshpande, V., et al., Eur. J. Biochem. 90, 191-198 (1978).

6. Wood, T.M. and S.I. McCrae, Biochem. J., 171, 61-72 (1978).

7. Ayers, A.R., et al., Eur. J. Biochem. 90, 171-181 (1978).

8. Wood, T.M. and S. I. McCrae, Proc. Bioconversion Symp., IIT Delhi, 111-141 (1977).

9. Reese, E.T., et al., J. Bacteriol. 59, 485 (1950).

10. Selby, I. and C.C. Maitland, Biochem. J. 104, 716-724 (1967).

11. Li, L.H., et al, Arch. Biochem. Biophys 111, 439-447 (1965).

12. Wood, T.M., Proc. IV IFS: Ferment. Technol. Today, 711-725 (1972).

13. Nisizawa, et al, " " " " 719-725 (1972).

14. Okada, G., J. Biochem. 72, 33-42 (1975).

15. Gong, C.S., et al, Biotechnol. Bioeng. 19, 959-981 (1977).

16. Gong, C.S., et al. Biotechnol, Bioeng. 21, 167-171 (1979)

17. Fagerstam, L., et al., Proc. Bioconversion Symp., IIT Delhi, 165-178 (1977).

18. Wood, T.M., Biotechnol. Bioeng. Symp. No. 5, 111 (1975).

19. Berghem, L.E.R. and L. Göran Pettersson, Eur. J. Biochem. 37, 21-30 (1973).

20. Berghem, L.E.R., and L. Göran Pettersson, Eur. J. Biochem., 46, 295-305 (1974).

21. Berghem, L.E.R., et al. Eur. J. Biochem, 61, 621-630 (1976).

22. Gum, E.K. and R.D. Brown, Biochim. Biophys. Acta, 446, 371-386 (1976).

23. Shoemaker, S.P. and R.D. Brown, J. Biochim. Biophys, Acta, 523 133-146 (1978).

24. Shoemaker, S.P. and R.D. Brown, Jr. Biochim. Biophys. Acta, 523, 147-161 (1978).

25. Sasaki, T., et al. Biotechnol. Bioeng. 21, 1031-1042 (1979).

26. Fan, L.T., et al., Biotechnol. Bioeng., 22, 177-199 (1980).

27. Ghose, T.K. and V.S. Bisaria, Biotechnol. Bioeng. 21, 131-146 (1979).

28. Huang, A.A., Biotechnol. Bioeng. 17, 1421-1433 (1975).

29. Okazaki, M. and M. Moo-Young, Biotechnol. Bioeng., 20, 637-663 (1978).

30. Howell, J.A. and J.D. Stuck, Biotechnol. Bioeng., 17, 873-893 (1975).

31. Howell, J.A. and M. Margat, Biotechnol. Bioeng., 20, 842-863 (1978).

32. B.S. Montencourt and D.E. Eveleigh, Proceedings of Second Annual Symposium on Fuels from Biomass, Vol. II, p. 613 (June 20, 1978) Rensselaer Polytechnic Inst., Troy, N.Y.

II. Production of Cellulase and Xylanase

A. Cellulase Production

1. Introduction

The overall conversion of biomass to ethanol through the enzymatic conversion (1,2,3) of the cellulose to glucose and the subsequent fermentation of the glucose syrups to ethanol has been hampered by two economic bottlenecks; the high cost associated with delignification and with enzyme production (4,5,). A doubling of cellulase production is possible by increasing the cellulose concentration (6,7) in the medium from 2.5 to 5.0%, increasing the nitrogen concentration, and controlling pH during growth (8). Culture filtrates from 2.5% cellulose cultures can reduce the hydrolysis time on a practical saccharification to one-half that required by culture filtrates from 1.0% cellulose cultures.

2. Materials and Methods

a) Inoculum

Viable cultures of *Trichoderma reesei* Rut C-30 and QM-9414 were maintained on PDA slants. The inoculum build up shown in Table II-1 resulted in faster rate of enzyme production and decrease in foam formed during the fermentation.

b) Production of Medium

A variety of media was employed, all based on the recipe reported by Mandels (6), and the modifications are listed in Table II- 2.

c) Fermentor System

Batch fermentationswere carried out in a 14-liter fermentor (New Brunswick Magnaferm Model MA 114) with an operating volume of 10 liters and all the system facilities for control of pH, temperature, agitation, dissolved oxygen and foam. The dissolved oxygen was automatically controlled at a level greater than 20%

Table II-1

Inoculum Build-up of T. reesei for a 5 or 14 liter fermentor

PDA SLANT CULTURE

↓

Incubated 200 ml mineral salts medium [containing 1% glucose, Tween-80 (0.01%) and antifoam (0.1%) in 500 ml Erlenmeyer flasks for 72 hours on an orbital shaker at 28°C, which was previously inoculated with above slant culture.

↓

Incubated 200 ml mineral salts medium [containing 1% Solka Floc, Tween-80 (0.01%) and antifoam (0.1%)] in 500 ml Erlenmyer flasks for 96 hours on an orbital shaker at 28°C, which was previously inoculated with above medium.

↓

fermentor

Table II-2

Composition of Media

Component (g/L)	Cellulose Concentration (g/L)		
	10	25	50
$(NH_4)_2SO_4$	1.4	3.9	11.5
KH_2PO_4	2.0	2.0	3.6
$MgSO_4$	0.15	0.15	0.3
$CaCl_2 \cdot 2H_2O$	0.4	0.4	0.79
NH_2CONH_2*	0.3	0.3	0.57
Peptone	1.0	1.6	2.9
Tween-80	0.2	0.2	0.2
Trace elements (mg/L)			
$FeSO_4 \cdot 7H_2O$	5.0	5.0	5.0
$MnSO_4 \cdot H_2O$	1.6	1.6	1.6
$ZnSO_4 \cdot 7H_2O$	1.4	1.4	1.4
$CoCl_2$	2.0	2.0	2.0

*Urea was deleted from the media for Rut-C-30 strain unless otherwise mentioned.

of the saturation value for the medium, by varying the agitation rates in response to changes in the dissolved oxygen. In most cases "one-sided control" was used.

For continuous fermentation (Fig. II-1), the set up can be divided into three parts, i.d. feed system 1), fermentors itself 2) and storage tank 3). The medium was sterilized every seven days in a vessel F having total volume of 40 liters and working volume of 30 liters. From vessel F, the liquid was pumped from time to time into vessel F_o, where medium could be kept into the stirred condition. Thus F_o would act as feed reservoir for fermentor F_1 and F_2, respectively. Pumps P_1, P_3 and P_5 are the high speed pumps which recirculates the suspension at a flow rate necessary to maintain a homogenous suspension of the cellulose in the line. The pumps P_2, P_4, and P_6 are the low speed pumps which draw feed into fermentor F_1, suspension of mycelium and cellulose in broth into fermentor F_2 and again homogenous suspension of mycelium and cellulose in broth into the reservoir R at points Y_1, Y_2 and Y_3, respectively.

The flow rate of pumps was regulated by preadjusting the time period of the timers T_1, T_2, T_3 and T_4.

d) Extracellular Component (Assay)

The filter paper assay, as described by Mandels, et al. (9), was measured by the release of reducing sugar produced in 60 minutes from a mixture of 1 ml diluted enzyme, 1 ml acetate buffer and 50 mg Whatman No. 1 filter paper incubated at 50°C.

Carboxymethyl cellulose (CMC) was determined by the increase in reducing sugar in 30 minutes from a mixture of 0.1 ml diluted enzyme and 1 ml of a solution of 2.0% CMC (in acetate buffer) incubated at 50°C.

C_1-activity was measured by the release of reducing sugar produced in 24 hours from a mixture of 1 ml diluted enzyme, 1 ml acetate buffer and 50 mg Red Cross absorbent cotton, incubated at 50°C.

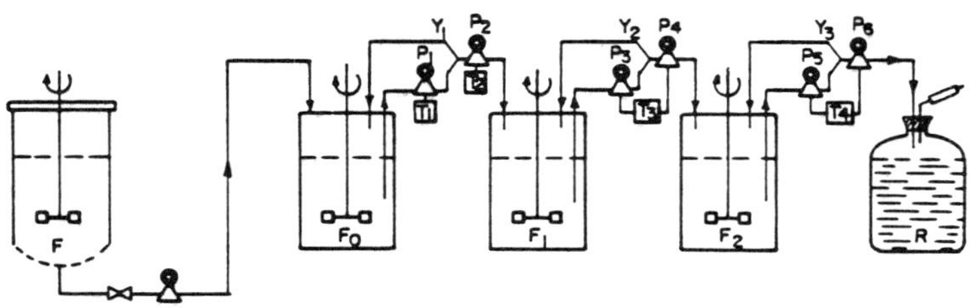

Schematic Diagram of Two Stage Single Stream Production of Cellulase

Figure II-1

β-glucosidase activity was measured by the release of glucose in 15 minutes as determined by the glucose oxidase-peroxidase assay, from a mixture of 0.1 ml diluted enzyme and 1 ml of 1.25% cellobisoe solution (in acetate buffer), incubated at 50°C.

Extracellular protein was estimated by the Lowery method (10)(without precipitation) using bovine serum albumin as the standard.

e) Cellular Components (Assay)

About 40 ml aliquot of the culture was filtered by suction through a tared 5 μM Nucleopore filter, washed with distilled water and dried at least overnight at 70°C, then weighed to measure the dry weight, which included mycelium and residual cellulose.

3. Evaluation of Trichoderma viride QM-9414

a) Batch Cellulase Production

Within 15 hours from the time of inoculation the pH of the media began to decrease. The dissolved oxygen reached at lowest, 20% saturation at 20 hours and pH reached the control point of 4.0 at 18 and 21 hours for 2.4 and 4.8 wt% cellulose cultures, respectively. For the initial 48 hours of fermentation temperature was kept at 30°C and while for rest of the fermentation time period it was 28°C. Similarly for the initial 24 hours of fermentation, pH was not allowed to fall below 4, then it was lowered to 2.9 and kept for 24 hours and finally after 48 hours of fermentation it was raised to 3.3 and was allowed to increase voluntarily. The objective of pH-temperature profiling for the initial 48 hours was to increase cell dry weight in conjunction with increase in enzyme activity. Cell dry weight could be increased by conducting fermentation and 30°C and pH 5.0 for the initial two days. But at pH 5.0, the rate of carbohydrate consumption is more than the cellulase demands, which can result in the repression of the enzyme system and hence lower enzyme yields. If fermentation after 2 days was conducted at 30°C instead of 28°C, it would result in the decrease of filter paper activity by about 30%.

In all the experiments dry weight rose initially as peptone was consumed and then decreased as the cellulose concentration reached a critical level. Basically, the batch process can be divided into two phases (Fig. II-2. In phase-I most of the cellulose is consumed with corresponding increase in cell dry weight. The cellulase activity is half of that of the total. During the rapid decrease in pH or the consumption of NH_4^+, saccharifying and endocellulase are induced. The rate of acid production is directly related to the rate of carbohydrate consumption. In phase-II, there is an increase in filter paper activity at the expense of decrease in cell weight and substrate exhaustion. Autolysis and sporulation are the main features of this phase.

b) Continuous Cellulase Production.

It was observed in previous work (11,12,13,14) that increasing the cell density or substrate concentration did not proportionally increase enzyme productivity. Extensive studies were carried out in order to optimize individually the 1st and 2nd stage of the two-stage continuous system for cellulase activity by manipulating pH, temperature, Tween-80 level, substrate concentration and dilution rates. The results are shown in Table II-3. The experiments were run continuously for about 3 1/2 months.

Runs #1 and #2 show that decreasing Tween 80 level by half increases the filter paper activity by 60% in the first stage of the two-stage fermentation. There is a considerable drop in the filter paper activity in the first stage at pH 5.0 than at other pH levels which severely effects the productivity in the second stage. Although, the filter paper activity of run #1 in the first stage is about 70% less than in run #3, the productivity is higher in

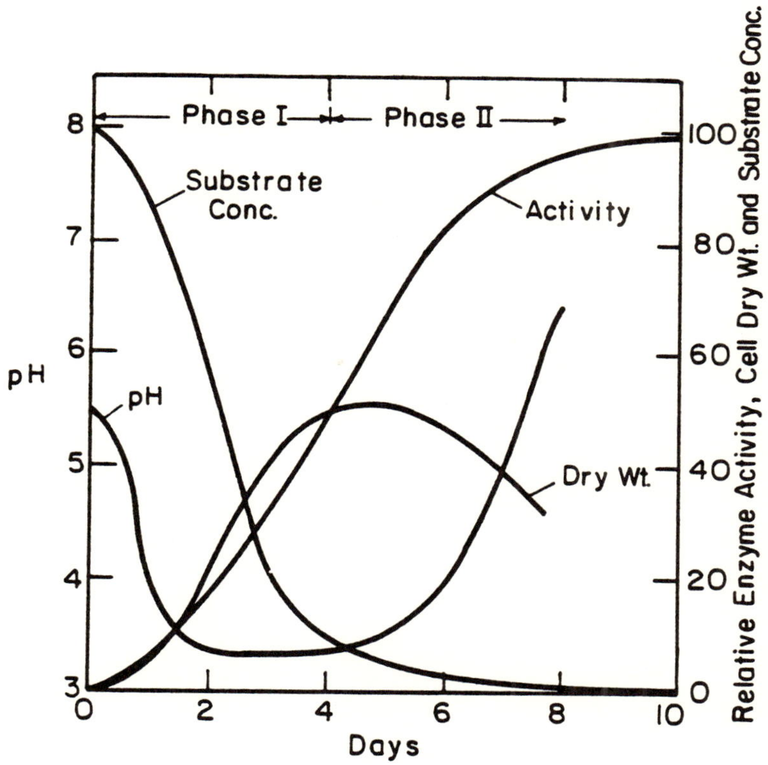

Growth and Enzyme Production of Trichoderma Viride on Cellulose

Figure II-2.

Table II-3
Two-Stage Continuous Production of Cellulase

Run #	CONTROLLED VARIABLES								RESULTS+					RESULTS(11) (PAST PROCESS)		
	Inlet Sub.** Conc. ($g\,\ell^{-1}$)	Tween 80 Level (%)	Temp (°C) F_1	Temp (°C) F_2	pH F_1	pH F_2	Dilution Rate (hr^{-1}) F_1	Dilution Rate (hr^{-1}) F_2	Productivity F_1	Productivity F_2	FPA F_1	FPA F_2	SYSTEM	S_o ($g.\ell^{-1}$)	D (hr^{-1})	FPA
1	25	0.2	28	28	4.0	5.0	0.02	0.02	0.053	0.112	2.64	5.6	Single Stage No Recycle	7.5 7.5 1.50 1.50	0.02 0.03 0.03 0.02	2.7 1.9 1.4 2.5
2	25	0.1	28	28	4.0	4.0	0.02	0.02	0.084	0.087	4.2	4.36				
3	25	0.1	28	28	3.75	3.3	0.02	0.02	0.089	0.105	4.44	5.24				
4	25	0.1	28	28	4.25	4.0	0.02	0.02	0.088	0.104	4.39	5.2	Single Stage Recycle	10	0.02	3.7
5	25	0.1	28	28	5.0	3.75	0.02	0.02	0.025	0.053	1.25	2.66				
6	25	0.1	28	28	4.0	5.0	0.027	0.027	0.054	0.116	2.0	4.3				
7	17.5	0.1	28	28	4.0	5.0	0.027	0.027	0.054	0.11	2.0	4.1				
8	17.5	0.1	28	28	4.0	5.0	0.054	0.054	0.038	0.099	0.7	1.84				
9	17.5*	0.1	28	28	4.0	4.00	0.02	0.02	0.05	0.096	2.54	4.8				

*Solka Floc (40 mesh)
**Solka Floc (200 mesh)

+For non-recycled system
F_1 = first stage ; F_2 = second stage

the second stage. If run #1 was operated with 0.1% Tween 80 level, it could lead to higher productivity. Similarly if the second stage of run #2 was operated at pH 5, it could also lead to higher productivity. Hence for all practical purposes the first stage can be operated between a pH range of 3.75-4.25 and the second stage at pH 5.0. In terms of inlet substrate concentration, 1.75% gives as good a filter paper activity as 2.5%. If the first stages of run #2 and run #9 are compared, it is seen that there is a decrease in filter paper activity and enzyme productivity from 4.2 and 0.084 to 2.54 and 0.053, respectively. If enzyme activity and production in the second stage are taken into consideration, then it would be profitable to use SW-40 rather than BW 200 as the substrate. Table II-4 gives the optimum operating conditions of the two-stage continuous cellulase production system using T. viride QM-9414.

4. Enhanced Production of Cellulase, and β-glucosidase by Rutgers C-30

a) Introduction

The production of cellulases and hemicellulases was studied with Trichoderma reesei Rut-C-30. This organism produced, together with high cellulase activities, considerable amount of xylanase and β-glucosidase.

Three cellulose concentrations (1,2.5 and 5.0%) were tested to determine the maximum levels of cellulase activity obtainable in submerged culture. Temperature -pH profiling to increase viable cell mass to maximum levels and thereby enhancing fermentor productivity at the higher substrate levels is discussed.

Effect of temperature, pH, Tween-80 concentration and substrate concentration on the rates of mycelial growth and extra cellulase enzyme production is also discussed.

b) Batch Fermentation

Trichoderma reesei (Rut-C-30) grown on cellulose is an excellent source of cellulase suitable for further process development studies.

Table II-4

Optimum Operating Conditions

Control Variables	1st Stage	2nd Stage
pH	3.75-4.25	5.0
Temperature (°C)	28°	28
Dilution Rate (hr^{-1})	0.02	0.027
Inlet Substrate (%) Conc.	1.75	

c) Effect of Tween 80 Concentration

Three concentrations (0.01, 0.02 and 0.1%) were studied for there effect on filter paper activity.

There was a slight difference in filter paper activity with 0.01 or 0.2% Tween 80 level, but once the concentration of Tween 80 was increased to 0.1%, there was a substantial decrease (40%) in filter paper activity. Similar effect was observed on other extracellular enzyme components. The mechanism of enhancement by Tween is not understood but may be related to: a) increased permeability of the cell membrane, allowing for more rapid secretion of the enzymes which in turn leads to greater enzyme synthesis of b) allowing the excess glucose to flow out of the cell into medium (thus relieving internal inhibition) because of the alteration of the transport phenomenon across the cell membrane. Tween 80 addition (0.1%) during saccharification increased the hydrolysis rate as well as glucose accumulation by about 30%.

This increase may be due to steady state concentration of cellobise kept in the hydrolyzate at a very low level.

Table II-5 gives a summary of the effects of control variables on state variables. It was observed by Wilke and Yang (8) that 31°C and a pH of 4.5 for the initial 48 hours and then 28°C and maintaining pH above 3.3 for the remainder of the fermentation time period was optimum for cellulase production.

In runs #1 and #2, pH and temperature were kept at 4 and 31°C, respectively for the initial 48 hours of fermentation. After 48 hours the pH was lowered to 3.3 and was controlled not to go below 3.3 while temperature was kept at 28°C and 25°C, respectively for 2 to 8 days. There appears not to be any appreciable difference in extracellular enzyme activities or soluble protein.

In runs #3 and #4, the pH was controlled not to go below 4 and 5.0, respectively. Temperature was controlled as in run #2. There is definitely

Table II-5

Effect of control variables on state variables

Operating Conditions						State Variables				
#	pH	Temp°C	S_o%	T.80%	C/N	FPA	β-glu.	C_1	C_x	S.P.
1	4 up to 48 hrs, after 48 hrs. decrease to	31° 0-2D 28° 2-8D	1.0	0.02	8.4	1.64	2.3	0.06	38	3.4
2	3.3 and was controlled not to go below 3.3	31° 0-2D 25° 2-8D	1.0	0.02	8.4	1.7	2.3	0.06	40	3.3
3	Controlled not to go below 4.0	31° 0-2D 25° 2-8D	1.0	0.02	8.4	2.0	3.55	0.169	44	3.4
4	Controlled not to go below 5.0	31° 0-2D 25° 2-8D	1.0	0.02	8.4	2.1	3.35	0.173	44	3.25
5	Controlled not to go below 5.0	31° 0-36H 25°RT	1.0	0.02	8.4	2.1	4.25	0.2	40	3.45
6	Controlled not to go below 5.0	31° 0-18H 25° RT	1.0	0.02	8.4	2.1	4.75	0.17	40	3.2
7	Controlled not to go below 5.0	31° 0-9H 25° RT	1.0	0.02	8.4	2.6	2.75	0.18	59.5	2.78
8	Controlled not to go below 5.0	25° 0-8D	1.0	0.02	8.4	3.1	3.3	0.195	84	3.3
9	Controlled not to go below 5.0	28° 0-8D	1.0	0.02	8.4	3.0	4.3	0.26	105	3.3
10	Controlled not to go below 5.0	28° 0-2D 25° 2-8D	1.0	0.02	8.4	2.9	3.9	0.23	115	3.45
11	Controlled not to go below 4.0	25° 0-8D	1.0	0.02	8.4	2.6	1.85	0.2	54	3.15
12	Controlled not to go below 4.0	25° 0-8D	1.0	0.02	8.4	2.1*	1.6	0.23	50	3.35
13	Controlled at 4.0	25° 0-6D	1.0	0.02	8.4	2.76	3.3	0.24	110	3.6
14	Controlled at 6.0	25° 0-8D	1.0	0.02	8.4	2.5	3.1	0.17	70	2.6
15	Controlled at 5.0	25° 0-8D	2.5	0.02	10.29	5.2	10	0.48	210	8.2
16	Controlled not to go below 5.0	25° 0-8D	5.0	0.02	8.1	14.35	26	1.03	348	20

* with urea (0.3 gm/l)

an increase in filter paper activity probably because of the increase in β-glucosidase as C_1 activities.

In runs #5 through 8, pH was controlled not to go below 5.0, and the initial temperature of 31°C was controlled for different intervals of time (36, 18, 9 and 0 hr.), while for the rest of the fermentation it was kept at 25°C. There is a definite increase in filter paper activity from 2.1 to 3.2 IU with a substantial increase in C_x (from 44 to 84 $IU.ml^{-1}$) activity.

In runs #9 and 10, the pH was controlled not to go below 5.0. The temperature was kept at 28°C for 0 through 8 days for run #9, while for run #10 it was kept for 0 through 2 days and then lowered to 25°C for the rest of the fermentation time period. There is not a substantial increase in filter paper activity, although the β-glucosidase and C_1 activities are higher.

In runs #11 and #12, the pH was not allowed to go below 4 and temperature was kept at 25°C. Moreover, in run #12, the effect of urea was studied. Addition of urea resulted in the decrease in filter paper and β-glucosidase activities.

In runs #13 and #14, the pH was controlled at 4 and 6.0 while the temperature was kept at 25°C. The filter paper activity at pH 4 (2.76 $IU.ml^{-1}$) is higher than at pH 6.0 (2.5 $IU.ml^{-1}$), but definitely less than when fermentation is conducted at pH $\geq$ 5.0 (run #8).

From the above observations it can be concluded that a temperature of 25°C and pH controlled not to go below 5.0 are optimum for enhanced cellulase production.

In runs #15 and #16, higher levels of cellulose (2.5 and 5.0%), respectively, were used. There is a substantial increase in cellulase activities as well as in soluble protein.

5. Comparison of Cellulases Derived from QM-9414 and Rutgers C-30

Table II-6 shows the comparison of Rut-C-30 with QM-9414. If we compare runs #2 and #3, the filter paper activity in #3 is slightly higher, but β-glucosidase activity is higher by about 9 times in run #2. This higher level of β-glucosidase would permit more rapid conversion of cellobise to glucose. This would then decrease the cellobisoe inhibition of the C_1 enzyme and hence increase the rate of depolymerization of crystalline cellulose.

If we compare runs #1 and #3 (Table II-5), there is an increase in filter paper activity, β-glucosidase, and soluble protein by about 3.7, 25.7 and 1.6 times, respectively.

All of these epxeriments demonstrate the superiority of Rut-C-30 over *Trichoderma viride* QM-9414.

6. Continuous Cellulase Production (Rutgers C-30)

Studies were carried out in order to optimize individually the 1st and 2nd stage of the two-stage continuous system for cellulase activity manipulating pH, temperature, Tween 80 level, substrate concentration and dilution rates. Part of the results are shown in Table II-7.

Run #1 and #2 show that decreasing Tween 80 level by half has no effect on filter paper activity. Increasing substrate concentration in the feed from 1% to 2.5% increases the filter paper activity from 1.03 and 1.6 to 2.48 and 3.6, respectively, in the first and second stage of the two-stage continuous system. Increasing dilution rate from 0.02 hr^{-1} to 0.04 hr^{-1} in the second stage results in the increase of productivity from 0.072 to 0.137.

B. Xylanase Production

1. Introduction

The utilization of biomass to provide a soruce of liquid fuels, such as ethanol, requires the conversion of all available sugars to ethanol. As most

Table II-6

Comparison of Rut-C-30 and QM-9414

Run #	S_0(%)	Strain	FPA U/ml	B-Glucosidase U/ml	Solution Protein mg/ml	Remarks
1	5.0	Rut-C-30	14.35	26	20	pH 5.0, T-80 level = 0.02%, 25°C
2	2.5	Rut-C-30	5.2	10	8.2	pH = 5.0, T-80 level = 0.02%, T = 25°C
3	5.0	QM-9414	6.06	1.01	12.68	(0-1Day)pH Allowed to fall to 4 (1-2Day)pH Allowed to fall to 2.8 (2D) raised to 3.3 and controlled not to go below pH 3.3
4	2.5	QM-9414	4.3	1.15	5.94	same as above

Table II-7

Two-State Continuous Production of Cellulase

Run #	CONTROLLED VARIABLES								RESULTS*			
	Inlet Sub** Conc. (g/l)	Tween 80 level (%)	Temperature (°C) F_1	Temperature (°C) F_2	pH F_1	pH F_2	Dilution Rate (hr^{-1}) F_1	Dilution Rate (hr^{-1}) F_2	Productivity F_1	Productivity F_2	FPA F_1	FPA F_2
1	10	0.02	25	25	5.0	5.0	0.02	0.02	0.021	0.032	1.03	1.6
2	10	0.01	25	25	5.0	5.0	0.02	0.02	0.020	0.030	0.98	1.49
3	25	0.02	25	25	5.0	5.0	0.02	0.02	0.050	0.072	2.48	3.6
4	25	0.02	25	25	5.0	5.0	0.02	0.04	0.052	0.137	2.6	3.42

* for non-recycled system
** ball milled solka floc (200 mesh)
F_1 = first stage; F_2 = 2nd stage

agricultural waste and woods contain 15-30% hemicellulose in the form of pentose polymers, it is clear that their depolymerization and subsequent fermentation is necessary to develop economical use of biomass. This depolymerization may be affected by mild acid hydrolysis or enzymatic hydrolysis. Acid hydrolysis may result in the formation of pentose degradation products, and involves later neutralization of the acid solution. For these reasons an enzymatic hydrolysis is preferable. The predominant pentose, xylose, so formed finds application in foods or confectionary, as an auxillary agent in the treatment of diabetes, and may be converted to alcohol, 1,4 butanediol or acetic acids.

Various molds, such as *Aspergillus foetidus* (15), *Asp. oryzae* (16), *Fusarium oxysporum* (17), *Trichoderma viride* (18), and *Chaetomium trilaterale* (19) have been reported as sources of xylanases. Some of these xylanases are reported to hydrolyze xylan to xylose, while other hydrolyze only xylobiose or xylo-oligosaccharides.

2. Materials and Methods

a) Microorganisms and Stock Culture

Streptomyces xylophagus nov. sp. selected as a xylanase producing bacterium was used. This strain was kindly supplied by Professor H. Iizuka at the Science University of Tokyo, Tokyo. Stock cultures were maintained on an agar-solidified mineral salts medium (Table II-8) but containing larchwood xylan at 0.5%.

b) Medium

The xylanase production medium (Table II-8) was the medium of Iizuka and Kawaminami (20). The medium was autoclaved at 250°F for 30 minutes to 1 hour, depending upon the liquid volume.

Table II-8

Xylanase Production Medium (20)

Constituents	Concentration (g/L)
xylan	10
K_2HPO_4	1
$MgSO_4 \cdot 7H_2O$	0.05
KCl	0.1
$FeSO_4$	0.01
Bacto-peptone	0.3-0.18
Tap water	to 1 liter

The inoculum was prepared by seeding 200 ml of fermentation medium with organism in 500 ml shaking flasks. This culture was incubated for 5 days at 30°C on an orbital shaker at 210 rpm and then added as inoculum (7.5% by volume) to the production vessel.

c) Fermentor

A standard 14-liter (New Brunswick) fermentor assembly provided with pH, temperature, agitation, dissolved oxygen and antifoam monitoring and control accessories was used for batch and continuous fermentation.

d) Xylan

The larchwood xylan used as substrate for the assay of xylanase activity was purchased from Sigma Chemicals. Prior to use, the xylan was subjected to the following treatment. A 1% suspension of xylan in de-ionized water was steamed for 10 minutes at 127°C. After cooling the suspension to room temperature, insoluble matter was removed by centrifugation or by filtration. To precipitate the xylan, 125 ml of ethyl alcohol were added to 100 ml of filtrate. The precipitate was collected by centrifugation, followed by evaporation of residual alcohol in a water bath, and lyophilized under vacuum to obtain the purified xylan.

The purified xylan was stored at 1.0% xylan in a mixture of 0.17 M KH_2PO_4 and 0.388 M NaOH. Prior to use, the 1.0% xylan solution was mixed with an equal volume of 0.35 M HCl so that the mixture of equal volumes produced a solution containing 0.5% xylan in 0.086 M phosphate buffer at pH 6.25.

e) Xylanase Activity (Assay)

For *Streptomyces xylophagus* 0.1 ml of enzyme solution (fermentation broth) was mixed with 0.5 ml (0.5%) xylan and incubated at 55°C for exactly 15 minutes. The reaction was stopped with the addition of 1.5 ml of DNS reagent and equivalent reducing sugar concentration was converted to total mg produced by 1 ml of the enzyme and recorded as the xylanase activity.

For Chaetomium trilaleral No. 2264 (Same as for Streptomyces xylophagus except incuation was at 50°C.)

Soluble protein was determined without precipitation using folin copper reagent described by Lowery (10).

3. Continuous Fermentation

The 14-liter fermentor with an operating volume of 10 liters was equipped with medium storage and product tanks. A volume of inoculum equal to 7.5% of the working volume of the fermentor was added and then allowed to grow as batch culture for 4 days to permit initial cell growth and enzyme production. After 4 days the feed of the fresh medium into the fermentor was started employing finger pumps (Sigma Motors), attached to the calibrated timer. A similar pumping arrangement was provided at the exit end of the fermentor to allow the homogenous suspension of cells in broth to flow out without any clogging. The temperature of cultivation was maintained at 30°C and the pH was automatically controlled at 7.4 by the addition of 2N NaOH or 2N H_2SO_4. The dissolved oxygen was automatically controlled at a level greater than 40% of the saturation value for the medium, by varying the agitation rates in response to changes in the dissolved oxygen. In most cases "one-sided control" was used.

a) Xylanase Production by Streptomyces xylophagus

In shake flasks, Streptomyces xylophagus nov. sp. grows in pellet form. The size of pellets varies from flask to flask, from very fine, with the largest about 1 mm in diameter. As pellets grow, the color of the broth changes from milky white to brown. The organism growing in wheat bran medium is thus difficult to observe directly.

The results of batch growth studies are shown in Table II-9. The enzyme production rate in shake flask and 14-liter fermentor showed a marked difference because of pH control in the fermentor. Similarly, there was a difference in

Table II-9

Xylanase Production

MODE OF OPERATION	TYPE OF SUBSTRATE	RESULTS
Shake Flask (200 ml medium in 500 ml flask)	Larchwood (L) 1% Wood Gum (WG)1% Wheat Bran (W) 7%	Final Enzyme Activity 6.97+ 6.89* 5.17 7.55 3.11++ 4.19** +Bacto-peptone 3g.ℓ^{-1} *Bacto-peptone 8g.ℓ^{-1} ++ Washed & Dried Wheat Bran ** Washed wheat bran
Submerged Fermentation Fer vol=14ℓ Liq vol=10ℓ	L ; WG ; W 1% 1% 7%	L: 9.82* WG: 8.53*(1.61)+ ; pH, 7.4 6.08 (1.75) pH, 7.9 6.83 (1.79) pH, 8.4 7.77 (1.81) pH, Uncontrolled 7.83 (1.56) pH, 8.4 Bacto-pep-2g.ℓ^{-1} 2.79 pH 8.4, No Bacto-peptone W: 9.44(3.16+) *Enzyme Activity (mg.ml^{-1}) +Soluble Protein (mg.ml^{-1})
Continuous (In 14 litre Fermenter)	WG(1%)	Dilution rate (hr^{-1}) — Steady State Value 0.02 — 6.75*(1.6)** (5.25)*** 0.027 — 7.25 (1.75) (6.5 0.034 — 5.7 (1.5) (7.25) 0.04 — 3.25 (1.4) (5.5) *Enz.Act(x.u) **Soluble protein (mg ml^{-1}) *** Cell Dry wt.(g.ℓ^{-1})

final enzyme activity in using washed-wheat bran compared with washed-dried wheat bran. In the 14-liter fermentor where the pH was controlled, wheat bran shows a considerable improvement in enzyme activity as compared to shake flask studies. A continuous culture dilution rate of 0.027 hr^{-1} gives an enzyme activity of 2.25 $mg.ml^{-1}$ while lower or higher dilution rates seem to effect the enzyme activity, soluble protein and cell dry weight. It thus appears that a dilution rate of approximately 0.027 hr^{-1} is optimal.

b) Xylanase Production by *Chaetomium trilaterale* No. 2264.

In shake flask, *Chaetomium trilaterale* grows in pellet form. The size of pellets varies from flask to flask, some being very fine and the largest is about 5 mm in diameter. As the pellets grow, the color of the broth changes from milky white to pink.

The results of batch growth studies are shown in Table II-10. The enzyme production rate in shake flask and 5-liter fermentor showed a marked difference because of pH control in the fermentor.

Table II-10

Xylanase Production

Mode of Operation	Type of Substrate	Results			Remarks
		Final Enzyme Activity			
Shake Flask (200 ml medium in 500 ml flask)	larchwood	12.7[a]	23.4[b]	4.69[c]	a,b, : with 1%, 2% substrate (0.1% Yeast) c , 2% Substrate, (0.05% Yeast)
	wheat bran	32.6[d]	32.5[e]		d,e: with 3.5%, 7.0% substrate, (0.1%, 0.2% yeast extract for 3.5% and 7.0% substrate conc.)
	larchwood	6.79[f]	4.45[g]	14.64[h]	Activities with 2,4 and 6 day inoculum age (f,g,h).

Mode of Operation	Type of Substrate	Results	pH	S_0(%)	Yeast Ext.(%)
Submerged fermentation	larchwood	0.54* (0.65)+	6.0	1	0.1
		6.60 (0.56)	7.0	1	0.1
Fer. vol: 5 l		1.95 (0.50)	6.5	1	0.1
		6.03 (1.09)	7.5	1	0.1
Liq. vol: 3 l.		1.95 (0.71)	8.0	1	0.1
		6.60 (1.28)	7.0	1	0.3
		6.25 (0.80)	7.0	1	0.2

*= Enzyme Activity(I.U./ml)
+= Soluble Protein (mg/ml.)

References

1. C.R. Wilke and G. Mitra, "Process Development Studies on Enzymatic Hydrolysis of Cellulose," Biotechnol. Bioeng. Symp. #5, 254 (1974).
2. C.R. Wilke and R.D. Yang, Proceedings of Symposium on Enzymatic Hydrolysis of Cellulose, Aulanko, Finland, 485 (1975).
3. C.R. Wilke and R.D. Yang, Applied Polymer Symposium, 175 (1975).
4. J.M. Nystrom and A. Allen, Biotechnol. Bioeng. Symp. 6, 55 (1976).
5. J.M. Nystrom and K. Kornuta, Proceedings Symp. on Enzymatic Hydrolysis of Cellulose, Aulanko, Finland, 181, (1975).
6. R. E. Andreotti, M. Mandels and C. Roche, Proceedings of Bioconversion Symp., I.I.T. Delhi, 249 (1977).
7. D. Sternberg, Biotechnol. Bioeng., 18, 1751 (1976b).
8. C.R. Wilke and H.W. Blanch, LBL-8658 (Dec 1978).
9. M. Mandels, R.E. Andreotti, and C. Roche, Biotechnol. Bioeng. Symp. #6, 21 (1976).
10. O.M. Lowery, N.J. Rosebrough, A.L. Farr, R.J. Randall, J. Biol. Chem., 193, 265 (1951).
11. C.R. Wilke, U. von Stockar and R.D. Yang., LBL-5267 (1976); AIChE Symp. Series 72, 104 (1977).
12. Communication, P. Nicolai, U.S. Army Natick R&D Command, Natick, MA.
13. M. Mandels, D. Sternberg and R.E. Andreotti, Symp. on Enzymatic Hydrolysis of Cellulose, Helsinki (1975) p. 81.
14. V. Sahai and T.K, Ghose, Proc. Bioconversion Symp. I.I.T. Delhi, P. 269 (1977).
15. R.L. Whistler and E. Masak, J. Am. Chem. Soc., 77, 1241 (1955).
16. S. Iskui, J. Gen. Appl. Microbiol., 4, 39 (1968).
17. H. Lyr and E. Novak, Zeit. Allg. Mikrobiol., 2, 86 (1961).
18. K. Nomura, T. Yasui, S. Kiyoka and T. Kobayashi, J. Ferment. Technol., 46, 634 (1968).
19. T. Kamaminami and H. Izuka, J. Ferment. Technol. 48, 161 (1970).
20. T. Kamaminami and H. Iizuka, Agr. Biol. Chem., 33, 1787 (1969).

III. Hydrolysis of Agricultural Residues

Of the many sources of biomass, the agricultural residues comprise a substantial resource. In the United States the straw residues of wheat, barley, corn, and sorghum average about 340 million metric tons yearly (1). It is obvious that with these amounts available containing an average of at least 50% carbohydrate and from 10 to 20% lignin, conversion of residues to sugars and other chemicals should be considered.

A. The Typical Conversion of Enzymatic Hydrolysis on Natural State Residues

A comprehensive study of various cellulose sources ranging from wood to agricultural residues performed by Andren *et al*. at the U.S. Army Natick Laboratories (2) showed that with relatively weak solutions of cellulase enzyme these materials could be hydrolyzed to sugars. For example, on rice waste (hulls and straw), the carbohydrate conversion was about 28%, and about 68% for material which had been ball milled to -200 mesh. Their results with corn residue, from the canning of sweet corn, consisting of stalks, cobs, leaves, and husks resulted in an exceptionally high conversion of about 53%. The usual carbohydrate conversions appear to be in the range of 25 to 40%.

In a study performed at this laboratory (3) several agricultural residues generated in California were systematically analyzed and enzymatically hydrolized to sugars. Additionally, a series of pretreatments were studied to see if the sugar production by enzymatic hydrolysis might be improved. Ultimately these studies were based as a basis of generating the parameters and process design for the Berkeley Process (4). The agricultural residues were the straws of barley, rice, sorghum, wheat, and corn stover. In addition rice hulls and cotton gin trash were included since the amounts of each are appreciable in California and they present environmental disposal problems.

Briefly, the experimental details of these bench scale studies involved 5 w/w% suspension of the appropriate substrate in cellulase enzyme solution made up to usually a total mixture of 250 grams. The pH optimum sppears to be between 4.7 to 5.3 and since it is somewhat broad, the cheaper acetate buffer instead of citrate buffer was used. All of the hdydrolyses were run at a pH of 5.0 (0.05 M acetate). The mixtures were stirred at about 180 RPM, which prevented 2 mm M Wiley milled particles from settling and gave good mixing, in a constant temperature bath at 45°C for 40 hours. Higher temperatures to 52°C are possible with apparent shorter hydrolysis time but at the expense of enzyme activity due to denaturation of the protein and consequent decrease in enzyme recovery. Lower temperatures only lengthen hydrolysis reaction times without a significant increase in enzyme recovery. More on this subject is mentioned in section III-D.

Since generally 95% of the ultimate carbohydrate conversion observed with most of these residues were obtained in 40 hours, the hydrolysis on all of the residues were reported consistent to that time.

The cellulose enzyme used in this study was derived from *T. viride* QM-9414 grown on N.F. solka floc SW-40 (80% sulfite treated and 20% Kraft Spruce, Brown & Company, Berlin, N.H.) and had a filter paper activity of 3.8, equivalent to 0.19 I.U. of glucose and 0.68 I.U. of cellobiose produced per ml. of enzyme. 27 liters of this batch of enzyme solution were stored frozen in 4 liter batches. Current use batches were stored at 3°C and preserved with 50 parts per million merthiolate to reduce bacteria growth which otherwise consumed the sugar as it was produced. This problem disappeared when the substrates were steam or chemically pretreated.

The products and reactants were usually separated by filtration or centrifugation. The liquid products were analyzed by gas chromatography and and the solids analyzed for carbohydrates and lignin composition for material

Table III-1

Yield of Sugars from Enzyme Hydrolysis (5w% Suspension) of Original Material
Basis: 100 lb. of Original Material Wiley Milled

MATERIAL	GLUC.*	POLY GLUC.	XYL.	ARAB.	G&P CONV. (%)	PENTOSE CONV. (%)	TOTAL SUGAR CONVERSION (%)
Barley	7.0	0.8	3.05	0.81	18.8	17.9	17.7
Corn Stover	11.2	0.5	3.01	0.62	30.1	25.7	26.4
Cotton Gin Trash	5.3	0.5	0.03	0.01	29.2	0.6	20.1
Rice Hulls	5.3	0.3	0.33	0.08	15.8	2.5	10.9
Rice Straw	17.5	0.01	2.55	0.84	42.7	17.6	33.4
Sorghum Straw	10.5	0.5	1.08	0.51	30.5	7.8	21.9
Wheat Straw	8.9	0.01	2.45	0.56	24.4	15.7	19.4

*Abbreviations for the sugars; polyglucose is mainly cellobiose.

balance.

The results of typical batch 40 hr hydrolysis with 5 wt% suspension the indicated agricultural residues, reported in (3), are shown in Table III-1. The only physical treatment, preceding the hydrolysis, was 2 mm Wiley milling. As can be seen, the total sugar production is not favorable.

B. Effect of Physical Treatments on Substrate for Enhanced Carbohydrate Conversion.

The two major physical treatments which give significant conversion enhancement are ball and roll milling. Such milling reduces the crystallinity of the cellulose, increases the surface area and bulk density, decreases the lignin sheath and thus makes the carbohydrate more accessible to the enzyme (2). Ball milling alone, or before and after chemical pretreatments, apparently seems to be the predominant factor for enhanced carbohydrate conversion observed in this laboratory and elsewhere (7,8). Newsprint ball milled to -200 mesh (<75μM particles) resulted in at least 84% carbohydrate conversion vs. untreated newsprint at about 25% conversion. Samples of wheat straw that were delignified by the common analytical method with chlorine dioxide/acetic acid and then followed by enzymatic hydrolysis in 5 wt% suspension resulted in carbohydrate conversion increase from 25% to 38%. Ball milling this delignified substrate to <105 μM particles increased the enzymatic conversion to 67% (8).

It would appear that any process whereby the crystallinity of the carbohydrate is reduced is the predominant factor for high carbohydrate to sugar conversions. This may be accomplished by physical pulverization which produces lower degrees of polymerization of both the lignin sheath and the carbohydrate.

C. Chemical Pretreatment

It has been shown (9) there are two modes of swelling for cellulose, intercrystalline and intracrystalline. Intercrystalline swelling in cellulose with water is obtained by entry of water molecules between crystalline units with an increase in volume equivalent to the volume of water adsorbed. Upon drying, the substrate reverts in time to its original dimensions and structure. Obviously, permanent disorder of the cellulose structure and hence the crystallinity is not accomplished by boiling with water.

Intracrystalline swelling involves penetration of the crystalline regions with concomitant disorder of the crystalline structure. This is usually accomplished with hot or cool alkali solutions or with cool concentrated phosphoric acid. Unlimited swelling by solution is complete disorder which presumably is permanent while in solution. The term "amorphous cellulose" is used to describe precipitated cellulose in a highly disordered state. These mixed structures (so-called β, γ, or cellulose -II, III, IV in x-ray deffraction terminology) revert in time to the stable α or microcrystalline cellulose-I (9).

1. Alkali and Ammonia Swelling

Alkali such as sodium hydroxide and ammonia are examples of limited swelling agents that cause definite alterations in the native cellulose structure. The extent of the lignin sheaths also affects the extent and kinetics of the swelling since ether bonds of the lignin moeties must also be disrupted or broken.

The use of alkali pretreatments on agricultural residues are numerous. Historically, the prevalence of alkali treatment arose from their use toward upgrading the nutritive value of forage and forest residues for ruminants. This was accomplished at far lower concentrations of alkali than was needed for mercerization of cotton which appears to be the percursor or basis of most alkali pretreatments. It was also assumed that *in vitro* rumen digestion of treated straws were an indication of increased hydrolytic activity of the carbohydrate with enzymes. This was demonstrated with wheat straw treated with alkali (10).

Toyama and Ogawa (11,12,13) boiled bagass and rice straw in 1% sodium hydroxide solutions for 3 hours and then enzymatically saccharified these treated substrates with cellulose enzyme derived from T. viride. 17% sugar solutions were reported for hydrolysis of 25 w% suspensions of alkali treated rice straw. If rice straw contains an average of 56% carbohydrate (63% sugar equivalent) and about 20% lignin, then this represents about a 91% carbohydrate conversion assuming the alkali completely solubilized the lignin. There is evidence that the alkali does not solubilize lignin to this extent (3). Janus (14) reported that chemical pretreatment with alkali on rice hulls hardly increased protein yield on hydrolysis wereas ball milling increased cellulose decomposition by 50% of the original assay of 42% by weight.

In studies performed in this laboratory, agricultural residues were also subjected to an alkali pretreatment (3). The basis of this pretreatment was that reported by Toyama and Ogawa (12), whereby the residues were boiled with 1 w% sodium hydroxide for 3 hours in 6 to 8 w% suspensions. The results on the enzymatic hydrolysis are shown in Table III-2. Clearly the increased sugar conversion to 61% observed for rice straw is not sufficient to overcome the cost of alkali consumption shown in Table III-3. This study demonstrates the high consumption of alkali that occurs with this type of pretreatment. It should be noted that this should not be unexpected when the primary reason for alkali treatment is swelling or insertion of hydroxide between the crystal lattices of cellulose. It also follows that additional alkali would be consumed in depolymerization to solubilize the lignin. The consumption of alkali shown in Table III-3 is the minimum expected as the agricultural residues were previously acid extracted, neutralized, washed and then dried.

Table III-2

Yield of Sugars from Enzyme Hydrolysis
(5 w% Suspension) of Acid and Base Treated Solid
Basis: 100 lb. of Original Material

MATERIAL	GLUCOSE	POLY GLUCOSE	XYL.*	ARAB*	CONVERSION OF AVAILABLE CARBOHYDRATE (&) G & PG	PENTOSE	SUGAR
Barley	16.0	2.4	0.21	0.06	47.7	2.2	36.7
Corn Stover	13.8	8.1	0.93	0.07	66.4	35.7	64.0
Cotton Gin Trash	6.9	1.8	0.32	0.05	44.8	6.1	35.6
Rice Hulls	7.5	1.2	0.01	0.01	27.2	0.3	21.7
Rice Straw	15.6	8.3	0.55	0.16	66.9	15.8	60.8
Sorghum Straw	13.9	9.7	1.8	0.23	80.8	22.8	67.3
Wheat Straw	14.6	3.5	2.07	0.10	57.5	36.5	53.4

(a) Acid treated solid extracted for 3 hours at 100°C with 1w% (0.25$\underline{M}$) sodium hydroxide.

(*) Including small amounts of polymeric pentose.

Table III-3

Alkali Treatment on Previously Acid Extracted Materials.

MATERIAL	LBS. NAOH USED/LB. GLUCOSE PRODUCED	ALKALI USED (%)
Barley Straw	0.341	55.2
Corn Stover	0.388	63.4
Cotton Gin Trash	0.696	50.2
Rice Hulls	0.484	43.9
Rice Straw	0.253	40.0
Sorghum Straw	0.347	59.9
Wheat Straw	0.320	57.6

Alkali with oxygen process employs an oxygen atmosphere over the alkali solutions. Though much research is being done for pulping on the industrial level because of the expectation of tightening of pollution standards, studies on this process have not been applied to agricultural residues.

Ammonia pretreatment is another long standing approach to up grading the nutritional value of forage residues (15). As with sodium hydroxide, the pertinent action of the ammonia is the hydrolysis of the ether cross links of lignin, thereby providing more ready access to the carbohydrate for the enzyme. In general, the enhancement of carbohydrate conversion by ammonia treatment is less than that observed with sodium hydroxide (16).

Corn stover that was pretreated with ammonia in this laboratory (17) indicated an approximately equal carbohydrate conversion by subsequent enzymatic hydrolysis compared to the acid pretreated material. The corn stover carbohydrate conversion was 61% versus a 26% for the untreated material. However, while the ammonia treatment was effective, it does not appear attractive because of the pressure requirements and recovery problems.

2. Dilute Acid Pentosan Extractions

The dilute acid pretreatment primarily for pentosan extraction is a common hydrolysis technique used many years in the pulping industry. The acid concentration and solid to liquid ratios are manifest. It appears that concentrations of 0.1 to 1% acid and solid loadings of 6 to 30 parts are the most common.

The landmark work of Saeman (18), Harris (19) and Root (20) on the kinetics of high temperature acid hydrolysis of biomass residues, especially wood, have established the basic parameters for the conversion to sugars. The work by Dunning and Lathrop (21) served as a basis of the acid pretreatment studies performed at this laboratory (3). The milled agricultural residues were boiled in 6 to 8 wt% suspensions in 0.09 molar sulfuric acid for generally five and one half hours. The results of these extractions are shown in Table III-4. The pentose concentrations and conversions clearly show that, except for cotton gin trash, the extractions are favorable. The treated solids were then subjected to enzymatic hydrolysis for 40 hours, at 45°C, with cellulase enzyme derived from Trichoderma viride QM-9414. These results are shown in Table III-5, and the overall combined carbohydrate conversions are shown in Table III-6. It would appear that corn stover and rice straw are the most amenable and cotton gin trash the least favorable for consideration as substrates for enzymatic production of sugars. It was also found in this study that the acid liquor could be recycled on an additional two batches of substrate before extraction became unfavorable.

Another dilute acid pretreatment that appears effective is the short contact time, high temperature process developed at Dartmouth College (22). The substrate in a form of a slurry was heated rapidly, mixed with dilute acid at the inlet of the reactor, held for a short time of about 12 seconds and then rapidly quenched. Since this pretreatment was for such a short time, degradation of the carbohydrate was claimed to be minimal. Various acid concentrations of .1 to 1% by weight were used on substates such as delignified wood, newsprint, and corn stover. Studies with reactor temperatures of 180 to 220°C were also made on the above substrates. It was claimed that upon enzymatic hydrolysis of the pretreated material that nearly quantitative conversions of glucose were obtained. Enzymatic hydrolysis, of their pretreated corn stover, done at this laboratory confirmed the 97% glucan to glucose conversion.

Table III-4

Acid Extraction Liquor of Original Material[(a)]
Basis: 100 lb. of Original Material

MATERIAL	GLUC.*	POLY GLUC.	XYL.	ARAB.	OTHER	PENTOSE CONV. (%)
Barley Straw	1.2	2.0	5.9	2.4		38.5
Corn Stover	2.9	1.0	12.2	2.4	3.3 Sol Lig	82.9
Cotton Trash	0.09	0.35	0.32	0.05	-	6.3
Rice Hulls	2.7	1.3	8.1	1.6	2.8 " "	56.7
Rice Straw	1.7	2.1	11.2	2.5	2.2 " "	70.1
Sorghum Straw	0.31	2.3	8.4	1.5	3.0 " "	50.5
Wheat Straw	2.5	1.2	10.9	2.0	--	67.6

(a) Extracted 5 1/2 hours (3 1/2 hours for corn stover) at 100°C with 0.92 w% (0.09<u>M</u>) sulfuric acid.

* Abbreviations for the sugars. Poly Glucose is mainly cellobiose. The small amounts of poly pentoses are included with xylose and arabinose.

Table III-5

Yield of Sugars from Enzyme Hydrolysis (5 w% Suspension) of Acid Treated Solid
Basis: 100 lb. Original Material

MATERIAL	GLUCOSE	POLYMERIC GLUCOSE	XYLOSE*	ARABINOSE*	Conversion of Available Carbohydrate (%) G&PG	PENT	SUGAR
Barley	13.0	0.5	0.32	0.06	34.6	2.9	26.7
Corn Stover	15.7	1.1	0.81	0.11	46.2	27.0	44.5
Cotton Gin Trash	5.7	0.8	0.59	0.06	33.3	10.2	27.6
Rice Hulls	4.0	0.8	0.04	0.03	14.9	0.0	12.3
Rice Straw	18.6	2.6	0.95	0.10	57.0	18.8	52.0
Sorghum Straw	14.7	0.3	0.82	0.13	44.8	9.0	36.3
Wheat Straw	12.5	1.2	2.36	0.10	41.8	38.4	41.2

* including small amounts of polymeric pentose

Table III-6

Total Yield Summary of Liquor and Enzyme Hydrolysis of Acid Treated Material
Basis: 100 lbs. Original Material

MATERIAL	GLUCOSE	POLY GLUCOSE	PENTOSES	G&PG CONVERSION (%)	PENTOSE CONVERSION (%)	TOTAL SUGAR CONVERSION (%)
Barley Straw	14.2	2.5	8.7	40.4	40.4	38.6
Corn Stover	17.6	2.1	15.5	50.8	86.3	60.6
Cotton Gin Trash	5.8	1.2	1.0	35.3	14.7	27.5
Rice Hulls	6.7	2.1	9.7	24.7	57.3	33.0
Rice Straw	20.3	4.7	14.8	61.6	76.6	63.7
Sorghum Straw	15.0	2.6	10.9	49.1	53.3	49.5
Wheat Straw	15.0	2.4	15.4	47.9	78.3	53.5

3. Nitric Oxide with Alkali

Although the results of acid pretreatment scheme are promising it does not accomplish an objective of delignification which is desirable in the enzyme production stage of the Berkeley Process. A processing scheme which meets this criterion is the nitrogen oxide process work in this laboratory by Borrevick, Wilke and Brink (23). In this process the wheat straw was first reacted with nitric oxide (a convenient way to reduce an air pollutant) and air at atmospheric pressure. The gas treated material is then extracted in water, and as in the dilute acid process the extraction liquor is rich in xylose. However, it also contains a sizable fraction of the lignin. An additional amount of lignin can be extracted into a dilute alkali solution representing an additional delignification. It was also found that a sequential mode of gas addition gave better results than simultaneous addition. When nitric oxide was first added to the reaction vessel and then a few minutes allowed to pass before the addition of oxygen (air), the reaction on the substrate appeared to occur more uniformly with consequent increased yield of carbohydrate conversion. The optimum condition of the gas phase reaction appeared to be 5 parts of nitric oxide per 100 parts of solid material, a 24 hour reaction time at 20°C and about 20% moisture content in the solid. In this way the overall conversion was about 60% for wheat straw. Sensitive analysis indicated that 42% of the pretreatment cost, excluding hydrolysis, was attributed to nitric oxide production. This presumably would be somewhat lower if the plant were situated near another plant that was releasing oxides of nitrogen that otherwise would be a pollutant.

4. Explosive Steam Decompression

Several years ago the Masonite Co. developed on a commercial scale, a process whereby wood chips were quickly heated with about 600 psi steam for about 10 minutes and just as quickly the pressure was released to atmospheric pressure by dumping the whole load into a large bin (24). In this manner the wood chips were essentially exploded into long wool-like fibers. The fibers were washed free of the pentose sugars and other water soluble constituents, and then they were fabricated into press board. The wash liquid was concentrated and sold as an animal feed supplement under the trade name "Masonex." The latter product contained 65% solids, of which 10% were simple sugars and 55% carbohydrates. The carbohydrate on hydrolysis was equivalent to 35% simple sugar with a distribution of 0.14 glucose, 0.27 mannose, 0.08 galactose, 0.05 arabinose, and 0.46 xylose.

Another process, similar to the above, receiving considerable attention is the Iotech process (25). Their research has shown that enzymatic saccharification yields of at least 60% can be achieved from tough ligno-cellulosic substrates, such as populus wood, with about 550 psi steam explosion treatment. The nature of the substrate dictates the severity of the steam treatment. The losses of the pentosans or conversion of the pentoses to furfural appears to be directly proportional to the cook time and/or the temperature. It was also claimed that the process produces reactive lignin that could be used as a thermoplastic binder.

5. Wet Oxidation Process

Recent research on a wet oxidation process has shown promising results (26,27). Wet oxidation in a batch process involves the treatment of aqueous slurry of biomass with air or oxygen at elevated temperatures and pressure through a two-stage system. In the first hydrolysis step the solid undergoes mild hydrolysis at pH $\sim$ 2, about 160°C, and 225 psi for the desired length of time. This is about 60 minutes for white fir. Samples are taken to monitor the course of the reaction for optimal sugar production versus time.

The reaction mixture is separated and the solid phase is then subjected to further oxidation at 180°C, or higher, in the second stage for optimal production of acids. The portion of the acid solution is then recycled to the next first stage containing a new batch of biomass.

In continuous operation, the solid used in the two hydrolysis stages flows countercurrent to the acid solution being generated from the wet oxidation of ligneous residue in stage three. In the first hydrolysis step the biomass undergoes mild hydrolysis as mentioned previously for the batch operation. The solids from the first stage are contacted with acid solution stream, generated in stage 3 and air is blown through the mixture at 170°C. In the second hydrolysis stage this solid mixture is heated to 200°C and reacted for about 35 minutes. The liquor from this stage is the product of the process containing the hemicellulose sugars, glucose, and organic acids, primarily acetic, formed mainly during the third stage.

By this processing scheme glucose yields of 45% have been claimed with a sugar concentration of about 4.6 w%. Though this process was developed for wood, it appears that little difficulty would be encountered in adapting it for use with agricultural residues that generally require less stringent conditions for carbohydrate conversion.

6. Ozone Pretreatment

An interesting pretreatment with ozone on biomass has been suggested (28). The main result claimed which might make the ozone treatment attractive is that the long chain structures of lignin and hemicellulose are cracked by ozonation. If this is true, then if follows that the cellulose chains are disrupted with at least lower degrees of polymerization. Another feature is that it sterilizes the substrate analogous to acid pretreatments.

D. Delignification Treatments

Much has been written about the efficacy of delignification, some of which appears factual and some of which appears to be in error. The error apparently arises when in a delignification the cellulose crystallinity and/or its degree of polymerization is also altered or affected. In such cases enhanced hydrolysis may be attributed to delignification when in fact it is due to alteration of crystallinity.

1. Chlorine Dioxide/Acetic Acid

As a means of comparison of various pretreatment effects, it is best to establish a standard for delignification. The procedures described by Moore and Johnson (30) employing sodium chlorite are acetic acid (29) are recommended. Warm acidified sodium chlorite liberates chlorine dioxide which in turns attacks lignin in biomass to form soluble products. The biomass carbohydrate is attacked only slightly under controlled conditions and can then be removed. This process is the basis of an analytical method for the determination of hollocellulose by delignification and has been used successfully on wood and annual plants for a number of years. Since the carbohydrate recovery is nearly quantitative it follows that if any changes occur, in addition to delignification, then it would be in crystallinity and not in the degree of polymerization. Decreases in the degree of polymerization or extensive changes in cellulose crystallinity to form the high disordered, the so-called "amorphous," form tends to increase the solubility of the cellulose.

The above method was used in a series of experiments in this laboratory originally to show the effect of delignification and the enhanced enzymatic hydrolysis of ground wood (31). The results clearly show the increase in saccharification as the original wood lignin content of 28.7% was decreased to 8.2%. About two years later the same ground wood sample was again enzymatically hydrolyzed and achieved only two-thirds as originally observed. The sample was reanalyzed and showed no apparent change in lignin or hollocellulose content, and the enzyme solution used was comparable to what was used previously. Since no other reasonable explanation was probable, it was concluded that the cellulose slowly recrystallized to the more ordered form, the so-called "α" or cellulose I, which has been shown to be resistant even to acid hydrolysis in a comparative study (32).

2. High Temperature Alcohol

The idea of using ethanol as a delignification solvent seems especially attractive because of its availability in the process in the fuels from biomass. In the work of Klinert (33) where 50 v% alcohol was used on wood at 180°C for about an hour, the lignin content was significantly reduced. This method was used (8) on samples of wheat straw at pH 6, and temperatures of 185 and 230°C, respectively. Upon subsequent enzymatic hydrolysis of the two samples, the overall yields of glucose plus cellobiose were 39.6 and 37.1%, respectively. The resultant lignin contents of the treated straws were 12.8 and 3.7% at 185°C and 230°C, respectively, compared to the original lignin content of 14.5%. Again, delignification in itself appears not to be sufficient to improve the enzymatic hydrolysis significantly. The dissolved lignin is easily recovered as a friable powder, so-called "amorphous" lignin, if the alcohol is removed by vacuum distillation at no greater temperature than 40°C. At temperatures of 45°C and higher the lignin converts to the thermoplastic, and apparently highly ordered, form. In the thermoplastic form, it appears to be an excellent binder and adhesive. The powdered form slowly converts to the thermoplastic form over a period of a year even when kept anhydrous and in vacuum.

3. Cadoxen

Cadoxen is a solution containing 5 to 7% cadium oxide in 38% aqueous ethylene diamine. This solution has been used to dissolve cellulose and from the determination of the resultant viscosity relate it to the degree of polymerization of the original cellulose. This concept was utilized (34) to dissolve Avicel and precipitate it by adding methanol and/or water followed by filtration. When enzymatically hydrolyzed the precipitated Avicel gave a 90% yield of glucose. The same process when applied to corn stover gave an 80% yield of glucose after only 3 hours of enzymatic hydrolysis. This study, although of dubious economic feasibility owing to the high cost of recovery and replacement of cadium and ethylenediamine, clearly shows that it is the crystalline structure of cellulose which controls the rate and extent of enzymatic hydrolysis.

E. Enzyme Recovery

As mentioned previously, an important consideration is to establish the optimum concentration of substrate for maximum carbohydrate conversion, minimum hydrolysis time and maximum enzyme recovery. The enzyme is strongly adsorbed on unhydrolyzed cellulose so that much of it is lost in the hydrolysis process. Using cellulase derived from *T. viride* QM-9494, and 5 w% substrate concentration (40 to 48 hour hydrolysis times with 20 to 30% carbohydrate conversion) about 50% of the original enzyme is released to the hydrolyzate and hence is potentially recoverable from solution. Enzyme release decreases even further as the substrate concentration is increased, a factor apparently not often considered in published work on enzymatic hydrolysis.

Work in this laboratory on untreated wheat straw and corn stover has shown with 5 to 6 w% substrate concentration, (40 hour hydrolysis) result in enzyme recoveries of about 50%. Following acid pretreatment of the above materials, the enzyme recovery on 40 hour hydrolysis approached 65%. Methods of enzyme desorption such as with phosphate ion gradient on cellulose and or with urea substitution (6) may permit greater enzyme recovery in the future.

More complete hydrolysis would also reduce enzyme losses since the enzyme is absorbed less strongly on noncarbohydrate material. Enzyme from the new mutant strain Rut-C-30 (Ref. 32, Section I) because of its much greater strength (Filter Paper Activity 14 I.U. per ml) shows great promise for more complete hydrolysis when this enzyme was used on corn stover and rice straw (5). The carbohydrate conversion was 1.5 times greater than that obtained on the same material with cellulase from *T. viride* QM-9414.

Considerable more work must be done to determine the parameters affecting hydrolysis kinetics, conversion and enzyme recovery and their interrelationships in order to establish economically optimum conditions for large scale processing.

References

1. G. H. Heichel, Amer. Sci. 64 70 (1976)
2. R.K. Andren, M. Mandels, J.E. Modeiros, Proc. Biochemistry, p.2-12, Oct. 1976.
3. A.F.Sciamanna, R.P. Freitas, and C.R. Wilke, "Composition and Utilization of Cellulose for Chemicals from Agricultural Residues," Univ. of Calif., Berkeley, LBL-5966, December 1977.
4. C. R. Wilke, R.D. Yang, A.F. Sciamanna, and R.P. Freitas, "Raw Materials Evaluation and Process Development Studies for Conversion of Biomass to Sugars and Ethanol," Univ. of Calif. LBL-7847, June 1978. Presented at and reported in -- Proc. 2nd. Ann. Symp. on Fuels from Biomass, Vol. I, p.421, June 1978, Rensselaer Poly. Inst., Troy, New York. To be published in Biotech. Bioeng.
5. J. Perez, and S. Wald, Private Communication, Univ. of Calif., Berkeley, Dept. of Chem. Eng., December 1979.
6. M. Riaz, and C.R. Wilke, "Studies of Desorption of Cellulase with Various Reagents," Univ. of Calif., Berkeley, LBL-7889, Oct. 1978.
7. M. Mandels, L. Hontz, J. Nystrom, Biotech. Bioeng. 16 , 1488 (1974).
8. C.R. Wilke, "Process Development Studies on Biocnversion of Cellulose and Production of Ethanol," Univ. of Calif., Berkeley, LBL-6860, June 1977.
9. J.A. Howsmon, and W.A. Sisson, "Cellulose and Cellulose Derivatives," Interscience Publ. Inc., New York (1954) p.17.
10. R.K. Wilson, and W.J. Pigden, Can. J. Anim. Sci. 44, 122 (1964).
11. N. Toyama, and K. Ogawa, Proc. IV Int. Ferm. Symp. Tech. Today, p. 743 (1972).
12. N. Toyama and K. Ogawa, Proc. N.S.F Seminar on Cellulose as Chemical and Energy Resource, Univ. of Calif., Berkeley, p. 375, June 25, 1974.
13. N. Toyama, and K. Ogawa, Proc. Bioconv. Symp. Indian Inst. Tech, Delhi, India, p. 373 (1977).

14. I.J.M. Janus, Proc. Bioconv. Symp., Indian Inst. Tech, Delhi, p. 469 (1977).

15. A.C. Weiss, et al., J. Anim. Sci. 35, 109 (1972).

16. H. Tarkow, W.C. Feist, Advances in Chem. Series 95 (1969).

17. C.R.Wilke,"Process Development Studies on the Bioconversion of Cellulose and Production of Ethanol," Univ. of Calif., Berkeley, LBL-6881, February 1978.

18. J.F. Saemen, Ind. Eng. Chem. 37, 43 (1945).

19. J.F. Harris, J.F. Saemen, E.J. Locke, "Chemistry of Wood," B.L. Browning, ed. Interscience Publishers, New York, 1963, Chapt. 11, p. 535-585.

21. J.W. Dunning, and E.C. Lathrop, Ind. Ing. Chem. 37, 24 (1945).

20. D.R. Root, J.F. Saemen and J.F. Harris, Forest Prod. J., p. 158, May 1959.

22. Diane R. Knappert, "Partial Acid Hydrolysis of Cellulose as a Means of Increasing the Rate of Enzymatic Hydrolysis., Masters Thesis, Dept. Chem. Eng. **Thayer** School of Engineering, Dartmouth College, June 1978.

23. R.K. Borrevik, C.R. Wilke and D.L. Brink, M.S. Thesis, "Effect of Nitrogen Oxide Pretreatments on Enzymatic Hydrolysis of Cellulose." Univ. of California Berkeley, Dept. of Chem. Eng., LBL-7879, September 1978.

24. D. Galloway, Pulp and Paper p.104, Aug. (1975).

25. Personal Communication, J. Noble, Iotech Corporation, 220 Laurier Avenue West, Ottawa, Ont., Canada.

26. L.L. Schaleger and D.L. Brink, TAPPI 61-4 65, 1978.

27. J.G. Bicho and D.L. Brink, " Wet Oxidation of Sokid Waste," Univ. of Calif. Sanitary Eng. Report. #71-1 pp. 1-86, Feb. 1971.

28. L. Pelloni, Brown-Boueri Co. Ltd., Zurich, CH8050, Switzerland.

29. L.E.Wise, Ind. Eng. Chem. Anal. Ed., 17 63 (1945).

30. W.E. Moore and D.B. Johnson, "Procedures for Chemical Analysis of Wood and Wood Products," U.S. Dept. of Agriculture, Forest Products Lab., Madison, Wisc. p. 6 Rev. Dec. (1967).

31. P.A. Carroad and C.R. Wilke,"Studies of Lignin-Degrading Fungi and Enzymatic Delifnification of Cellulosic Materials," Ph.D. Thesis, Univ. of Calif., Dept. Chemical Eng., Berkeley, LBL #4490, April (1976) p.67-72.

32. Ref. 19. p-137.

33 T.N. Kleinert, TAPPI 57, #8, 99 (1974).

34. G. Tsao, et al., Laboratory of Renewable Resources Engineering, Purdue University, Lafayette, Indiana, Feb. 1978).

IV. Enzymatic Hydrolysis Processes

A. Separate Hydrolysis and Fermentation

1. Stirred Tank Reactor Hydrolysis Processes (Natick, Berkeley, I.I.T.)

a) Process Description

The U.S. Army Natick Research and Development Command (Natick), the University of California (Department of Chemical Engineering and Lawrence Berkeley Laboratory) and the Indian Institute of Technology (IIT) have developed processes for bioconversion of cellulose to ethanol based on a stirred tank hydrolysis followed by a separate ethanol fermentation reactor. The basic steps for these processes are illustrated in Figures IV-1, IV-2, IV-3.

i) Pretreatment

Pretreatment of the cellulose is necessary to enhance the kinetics of hydrolysis by reducing the crystallinity and degree of polymerization of the cellulose and thereby making it more accessible to attack by enzyme. The Natick pretreatment consist of two roll compression milling and is assumed to render 75% of the urban waste feed cellulose hydrolyzable to glucose on subsequent hydrolysis (1). The Berkeley pretreatment consistsof shredding and hammermilling followed by dilute sulfuric acid treatment for corn stover and wheat straw. There is no acid treatment for newsprint. The acid treatment also extracts xylose which can be utilized for further ethanol production or can be considered a by-product (2). The IIT process utilizes ball milling such that highly concentrated (more than 30% cellulose) can presumably be handled in hydrolysis (3).

ii) Hydrolysis

The Natick hydrolysis of the cellulose fraction of urban waste is carried out semibatch over 24 hours on an initial substrate solids charge of 20%, which is increased to an effective 30% by additional charges in the

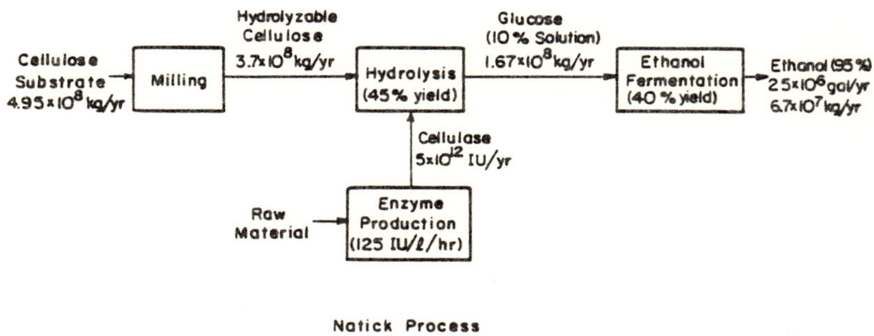
Cellulose Substrate 4.95×10⁸ kg/yr
Milling
Hydrolyzable Cellulose 3.7×10⁸ kg/yr
Hydrolysis (45% yield)
Glucose (10% Solution) 1.67×10⁸ kg/yr
Ethanol Fermentation (40% yield)
Ethanol (95%) 2.5×10⁶ gal/yr 6.7×10⁷ kg/yr
Cellulase 5×10¹² IU/yr
Raw Material
Enzyme Production (125 IU/ℓ/hr)
Natick Process

Figure IV-1.

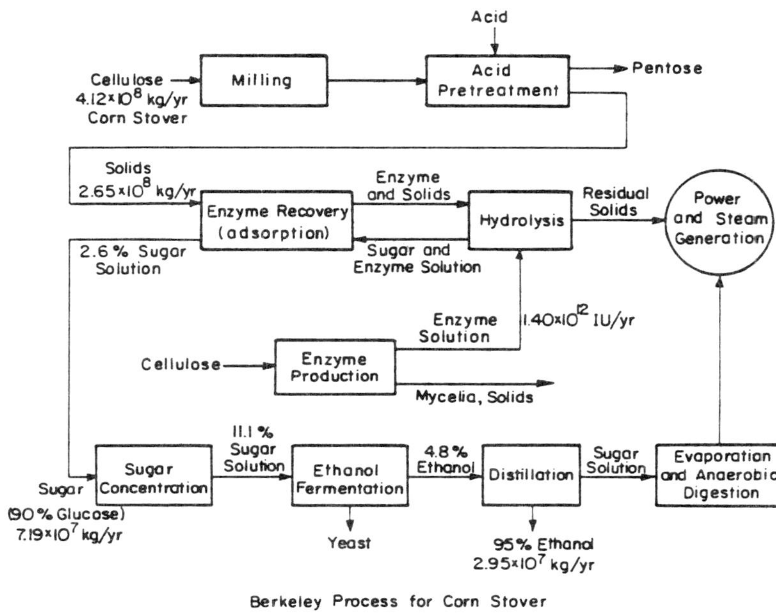
Acid
Cellulose
4.12×10⁸ kg/yr
Corn Stover
Milling
Acid
Pretreatment
Pentose
Solids
2.65×10⁸ kg/yr
Enzyme Recovery
(adsorption)
Enzyme
and Solids
Hydrolysis
Residual
Solids
Power
and Steam
Generation
2.6 % Sugar
Solution
Sugar and
Enzyme Solution
Enzyme
Solution
1.40×10¹² IU/yr
Cellulose
Enzyme
Production
Mycelia, Solids
11.1 %
Sugar
Solution
Sugar
Concentration
Sugar
(90 % Glucose)
7.19×10⁷ kg/yr
Ethanol
Fermentation
Yeast
4.8 %
Ethanol
Distillation
95 % Ethanol
2.95×10⁷ kg/yr
Sugar
Solution
Evaporation
and Anaerobic
Digestion
Berkeley Process for Corn Stover

Figure IV-2.

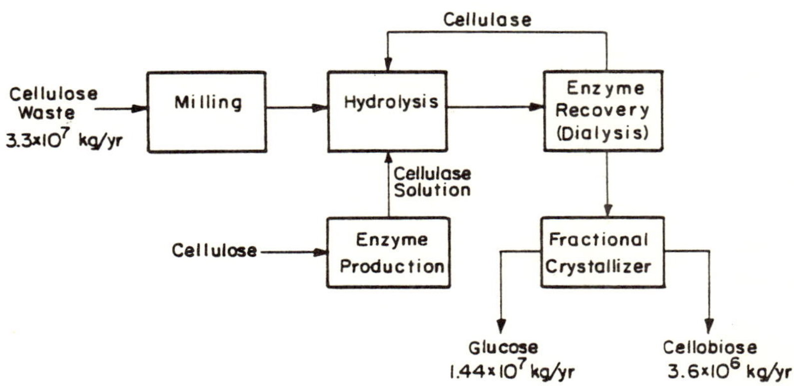

Indian Institute of Technology Process

Figure IV-3.

first few hours when the viscosity drops. Forty-five percent conversion of the hydrolyzable cellulose, which is 75% of the original cellulose feed, produces a 10% syrup of fermentable sugars (1).

The Berkeley hydrolysis is continuous for a solids feed of 5 wt% in agitated cylindrical concrete digestors at 45°C for over 40 hours. Forty percent of the potential glucose in the corn stover is assumed converted to glucose to produce a 2.6% sugar solution. This sugar concentration can be adjusted by varying the recycle of the sugar enzyme product stream (2).

IIT demonstrated a combined system for cellulose hydrolysis in a stirred tank reactor with glucose removal by a separate membrane cell. Using a semibatch operation over a ten day period, 71% of the Solka Floc feed was converted to glucose. About 17% of this glucose remained in the system without removal. The average mass flux through the membrane was dependent mainly on the solids concentration rather than the initial sugar concentration or the total amount of sugar removed. For a mixture of 14.06% sugar and 17.33% cellulose the average mass flux ranged from 2.9 to 8.4 gal/ft^2/day for molecular sieve membranes with molecular weight cut-offs of 10,000 and 30,000, respectively (4).

iii) Enzyme Production

The cellulase production processes for Natick, Berkeley, and IIT are all based on the aerobic fermentation of *Trichoderma reesei*. Natick has achieved an enzyme productivity of 125 IU/liter/hour in batch fermentation with recently developed hypercellulase producing mutant strains of this organism. One of these mutants is the Rutgers C-30 strain, which is catabolite repression resistant, i.e., its cellulase production is not repressed by glucose (5). Both the Berkeley and IIT processes are based on the QM-9414 strain. The Berkeley enzyme productivity is 19.2 IU/L-hr. for a continuous two-stage, growth and enzyme induction process. The cells from the second stage are recycled back to the first stage. Natick also recycles its mycelia to replace proteose peptone as a nitrogen source.

Berkeley and Natick currently use delignified cellulose as the carbon source for enzyme production. However, preliminary Natick studies indicate this pure cellulose substrate may possibly be replaced by urban waste in future processes.

iv) Enzyme Recovery

The Berkeley process recovers its enzyme by contacting countercurrently washed pretreated solids with enzyme sugar solution from the hydrolysis vessel in a two-stage mixer settler system. The overall enzyme recovery for this process is assumed to be 58% with most of the losses on the unreacted solids. Preliminary studies show that urea is effective in desorbing enzyme from cellulose and should lead to higher enzyme recovery. IIT also recovers enzyme from hydrolysis with a membrane cell which separates product sugar from enzyme and unreacted cellulose by a molecular sieve membrane. No recovery data for the enzyme was reported. As with the Berkeley process the degree of enzyme recovery will depend on the distribution coefficient of the enzyme between the liquid and solid substrate and also on the degree of enzyme deactivation. Natick does not consider enzyme recovery economical. One reason is that most of the enzyme may be trapped on the unreacted solids. Also the enzyme cost for the Natick process is relatively less expensive than for the Berkeley and IIT processes.

v) Sugar Concentration

In the IIT process, one part cellobiose is separated from four parts glucose by fractional crystallization after the sugars are separated by molecular sieve membranes. In the Berkeley process, the hydrolyzate sugars are concentrated from 2.6 to 11% with a multi-effect evaporator. The Natick hydrolysis provides 10% sugar solutions which do not require further concentration. The sugars from all the processes are then fermented to ethanol.

b) Process Evaluations

Natick originated the process concept of converting cellulose to glucose via enzymatic hydrolysis and has since played a major role in its development. Presently, Natick is operating a 125 lb/day pre-pilot plant with a 280 liter working volume enzyme production vessel and a 250 liter stirred tank hydrolysis reactor. The Berkeley and IIT processes have been demonstrated, thus far, only on a bench scale.

The major advantage of the Natick process relative to the others as described above is higher enzyme productivity. This is derived from the use of recently developed hypercellulase producing strains of *T. reesei*, e.g., the Rutgers C-30 mutant. These properties allow up to 30% solids to be converted to 10% sugar, which also eliminates the need for further concentration in the Natick process. However because of the high solids residue there is a problem of separating the sugar solution from the solids. If a wash is used, the sugar concentration would be diluted and require reconcentration.

The Berkeley process offers the advantages of continuous enzyme production, hydrolysis and enzyme recovery. With 58% enzyme recovery assumed, the Berkeley Process requires 21.9 I.U. of enzyme per gram of glucose produced. The Natick process with no enzyme recovery requires 29.9 I.U./gram glucose. The enzyme to glucose ratios indicate the continuous Berkeley process with enzyme recycle makes more efficient use of enzyme than the semibatch Natick process with no enzyme recovery. However, the difference between the substrates, corn stover and urban waste, and between the enzymes must also be considered in comparing the usage of enzyme in the two processes. The cost of the original enzyme and the cost of enzyme recovery are also important in assessing the advantages of enzyme recovery.

The Berkeley process utilizes much less expensive concrete digestors for hydrolysis reactors, but sterile operation of these digestors has been questioned (6). The absence of storage facilities for agricultural residues has also been questioned. Most importantly, the Berkeley process uses a low cellulose feed concentration to produce a low sugar concentration, which needs to be concentrated.

The IIT process is based on the same cellulose producing organism (QM-9414 strain of *T. reesei*) used in the development process for Berkeley and should have similar enzyme productivities. The IIT process also recovers enzyme. Even with enzyme recovery, the enzyme cost for sugar production for either Berkeley or IIT process should be higher than for the Natick process. Although the economics of the IIT process has not been reported, the membrane cost is expected to be high. Overall, the IIT process is considerably less developed than either the Natick or Berkeley process. It is still basically conceptual with little reported data on its performance.

2. Fluidized and Packed Bed Hydrolysis Processes

a) Fluidized Bed Hydrolysis

In addition to the use of conventional stirred tank reactors for hydrolysis, a cellulase-bead fluidized bed reactor is being developed to hydrolyze cellulose to glucose by the Tokyo Institute of Technology (7). The experimental apparatus for this process is shown in Figure IV-4.

The cellulase is immobilized by entrapment in a collagen fibril matrix supported on glass beads. No leakage of cellulase from the matrix was observed, and the stability of the entrapped enzyme with a half time of 21 days was greater than that of the native enzyme with a half time of only 30 hours. For a 0.33% cellulose (Avicel SP) suspension, the immobilized cellulase relative to native cellulase showed initially lower conversion but after 120 hours

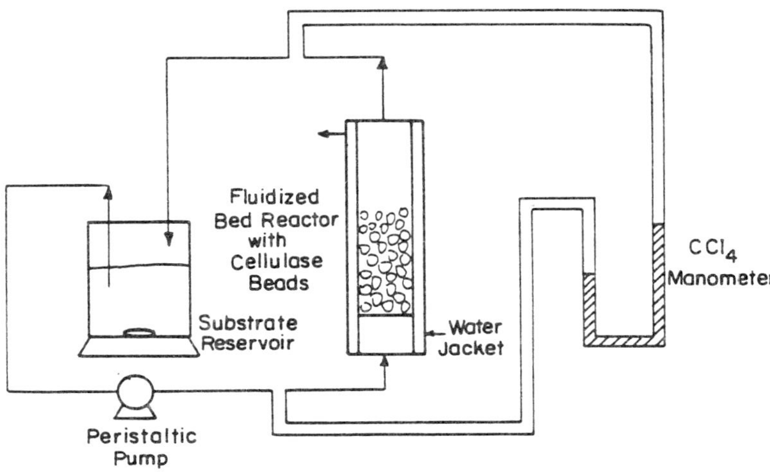

Tokyo Institute of Technology Experimental Apparatus for Cellulase-bead Fluidized Bed Reactor

Figure IV-4.

showed higher conversion, eventually giving quantative conversion. Therefore, the immobilized enzyme is advantageous for long reaction times. The pressure drop through the bed is low and not dependent on the flow velocity. The optimum flow velocity was 1 cm/sec.

A stable immobilized enzyme for cellulose hydrolysis has high potential for a practical continuous process in which the enzyme can be reused with minimal enzyme make-up. With a fluidized bed it is particularly easy to continuously replace portions of the bed. However, for the enzyme used by the Tokyo Institute of Technology, the feed cellulose concentration is too low and the reaction time too long to be economically feasible. Developmental work is continuing on scale-up of the reactor and use of waste cellulose.

b) Tapered Fluidized Bed Bioreactor

The Oak Ridge National Laboratory has developed a modified fluidized bed called the tapered fluidized bed bioreactor (8). Although this type of reactor has not been used for cellulose hydrolysis, it is suggested as a possible vehicle for fluidizing immobilized cellulase enzyme by Georgia Inst. of Technology (9).

The tapered bed is illustrated in Figure IV-5. The tapered bed is shaped like an inverted truncated cone with increasing diameter from the bottom entry point to the top exit point. This tapered configuration allows the feed at the bottom to be distributed equitably throughout the cross section without the back mixing which would occur with feed into a constant cross sectional area column. The decreased linear velocity at the top of the tapered bed also allows a higher flow rate range at the bottom without loss of materials from the top. Although this reactor has promise, its operating characteristics need to be understood better and its economic feasibility needs to be assessed.

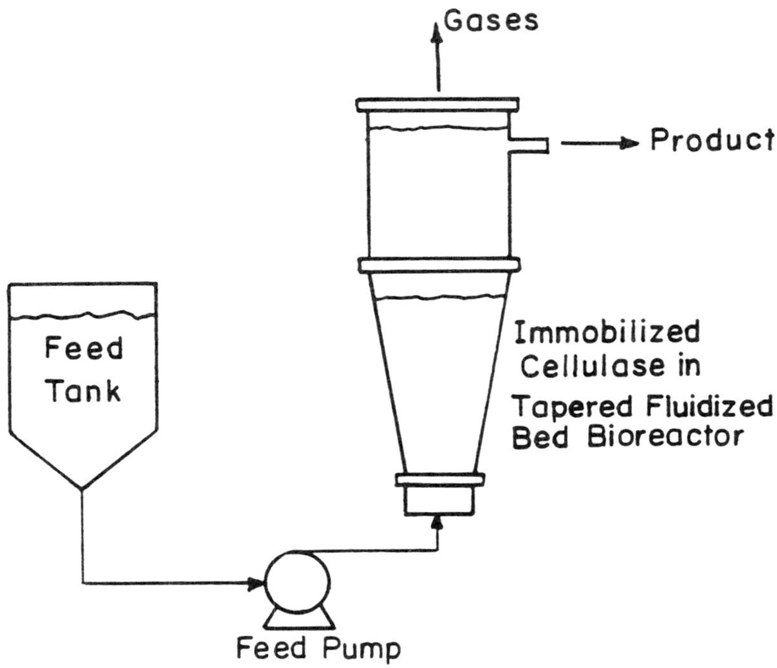

Tapered Fluidized Bed Bioreactor

Figure IV-5.

c. Fixed Packed Bed Hydrolysis

A fixed packed bed reactor design was also used by the Tokyo Inst. of Technology for cellulose hydrolysis (7). In this case the immobilized cellulase was still entrapped in the collagen fiber matrix but in a packed instead of a fluidized bed. They concluded that this packed bed configuration was not suitable for insoluble substrates. Major problems with the packed bed are plugging of the bed with a resulting high pressure drop in the bed, achieving uniform flow distripution, and replacing inactivated bed portions without removing the total bed (9).

Dynatech R/D Company, Cambridge Massachusetts also reported using a fixed packed bed fermentor for converting biomass, such as marine algae into organic acids (10). In this design, biomass is the stationary phase and nutrients and organisms are fed through the bed as the moving phase. This configuration allows a high solids loading, but M.I.T. found high pressure drops with this type of reactor and turned to fluidized beds (11). At present, it does not appear that either the immobilized enzyme packed bed or the packed cellulose bed are likely processes for cellulose hydrolysis.

3. Solid Culture Processes

Toyama, at Miyazaki University, has utilized various aspects of solid culture fermentation for enzymatic hydrolysis of cellulose separate from ethanol fermentation (12). These are production of cellulase by solid culture, hydrolysis with a solid culture of *T. reesei*, hydrolysis with a solid culture extract, and auto saccharification of solid cultures inoculated with *T. reesei*

a) Cellulase Production on Solid Cultures

In Japan industrial cellulase production is by growing *T. reesei* on solid cultures (Koji Method). The usual substrate is wheat bran, although rice straw and newspaper also produce high filter paper degrading activity enzyme. Delignified rice straw and delignified newspaper produced significantly less active filter paper degrading enzymes. Use of lactose in cellulase production was shown to inhibit the activity of the enzyme produced. The automatic Koji-making apparatus can be a stationary type, a rotary tray type, or a rotary drum type.

Using commercially prepared T. reesei cellulase on 25% delignified rice straw or bagasse substrate for 96 hours produces approximately 15 and 22% sugar syrups with 1 and 3% cellulase, respectively. Delignification was achieved by boiling with 1% sodium hydroxide solution for three hours. Although the sugar yields are high and the hydrolysis is fast, use of this commercially prepared enzyme is not considered ecomonically practical.

b) Hydrolysis with Solid Cultures

An alternative process would utilize T. reesei solid cultures grown on cellulosic waste for hydrolysis of cellulosic waste. For example, a rice straw solid culture was prepared by mixing rice straw and wheat germ in a 4:1 ratio. Two grams of this straw wheat germ Koji is then mixed with three grams of delignified rice straw (or bagasse) and suspended in 20 ml of acetate buffer of pH 5.0. The delignified rice straw and delignified bagasse produced 8.3% and 9.1% sugars, respectively, after 48 hours incubation. The disadvantage of a solid culture process is that considerable volume is occupied by non-convertible solids, which later also have to be separated.

c) Hydrolysis with Solid Culture Extracts

A more practical alternative process is for the solid culture, which was grown at 25-30°C for four days, to be extracted with water to produce enzyme solution with strong xylanase and cellulase activities. The enzyme solution then can hydrolyze the delignified cellulose. From five grams of delignified rice straw or bagasse treated with 20 ml of solid culture extract, 11.5% or 13.9% sugars are produced, respectively, after 48 hours. The residue after extraction contain conidia and mycelia of the fungus and has value as an antifungal agent. The solid culture extract process is shown in Figure IV-6 (12).

The high rate of hydrolysis and yields for hydrolysis of delignified rice straw and bagasse with rice straw solid culture extract recommends it as a high potential process. However, considerable developmental work is required for application of this laboratory scale process to industrial production. As with the Natick process and any other process in which there is a high sugar concentration with incomplete solids conversion, separation of the sugars from the residue is a problem.

d) Autosaccharification Process

The autosaccharification process consists of inoculating solid cultures of cellulosic waste with *T. reesei* and allowing the cellulase produced to hydrolyze the cellulosic waste to sugar. The major problem with this process is that using easily hydrolyzed substrates, such as delignified rice straw or delignified bagasse for cellulase production produce weak cellulase activity. Conversely, using difficult to hydrolyze substrates, such as wheat bran, rice straw, or bagasse produces high activity cellulase, but which is still unable to efficiently hydrolyze these substrates.

B. Simultaneous Hydrolysis and Fermentation

Hydrolysis of cellulose simultaneously combined with ethanol fermentation of the hydrolyzate sugars results in faster hydrolysis, higher sugar yields, and subsequently higher ethanol yields compared to separate hydrolysis and fermentation. The reason is that the end products of cellulose hydrolysis, mainly glucose and cellobiose, are usually inhibitory to the cellulase enzyme

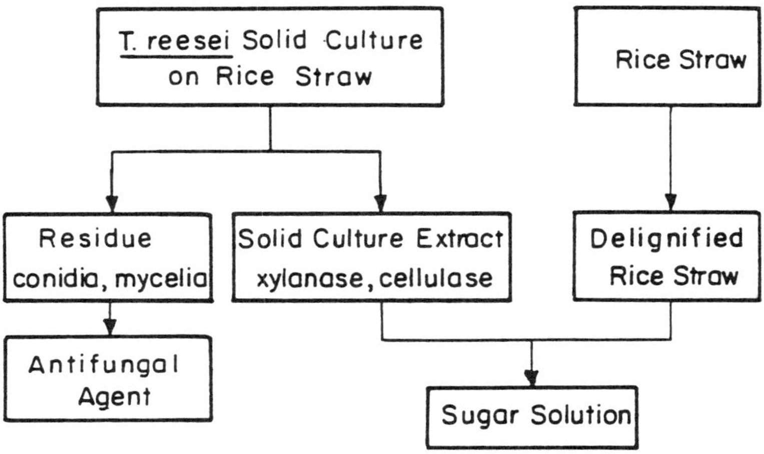

Miyazaki University Process for Hydrolysis with Solid Culture Extracts

Figure IV-6.

system. Depending on the yeast, either glucose or glucose and cellobiose are utilized in the fermentation. If cellobiose is utilized, its inhibition of cellobiohydrolase activity (C_1) is reduced. In either case glucose is removed by fermentation so that its inhibition of β-glucosidase, which splits cellobiose to glucose, is reduced. Thus, fermentation of glucose alone facilitates cellobiose removal.

The hydrolysis agent can be cellulase enzyme fed from an enzyme producing reactor or produced by an organism in the combined hydrolysis fermentation reactor. The latter might either hydrolyze cellulose to glucose and ferment the glucose to ethanol by itself, or be combined in a mixed culture with another organism, which will also ferment cellobiose, xylobiose, and xylose to ethanol and other by-products.

1. Enzyme Hydrolysis and Fermentation (Gulf/Arkansas Process)

a) Process Description

A process for simultaneous hydrolysis of cellulose with separate enzyme production and fermentation to ethanol was developed by the Gulf Oil Chemicals Company and later transferred to the University of Arkansas (13, 14). The basic steps of this process are shown in Figure IV-7. This process has been developed in a pilot plant utilizing one ton per day of cellulose at Gulf's Jayhawk plant in Pittsburg, Kansas. A 50 ton/day demonstration plant is presently being designed. A commercial 2000 ton/day plant which will produce about 50 million gallons of ethanol per year, is being planned for operation in 1983.

The feedstock is expected to consist of about 2/3 municipal solid waste (air classified fraction) and 1/3 pulp mill waste. This combined feedstock is assumed to contain 57% convertable cellulose. About 15% of the municipal solid waste is mechanically pretreated with a low-energy attritor,

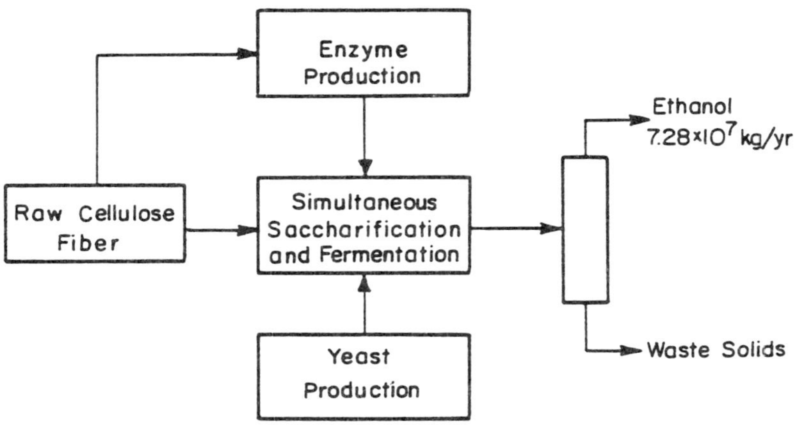

Gulf/University of Arkansas Process

Figure IV-7.

sterilized and then sent to the substrate to the enzyme reactors. The remaining municipal solid waste is mechanically pretreated, mixed with the pulp mill waste, pasteurized, and then becomes the substrate for the simultaneous saccharification and fermentation (SSF).

Using a mutant strain of *Trichoderma reesei*, cellulase is continuously produced in a submerged culture utilizing two trains of three reactors in series in each train. The total residence time for each train is 48 hours. The enzyme activity produced results in 90% utilization of the cellulose fraction of the solids fed to the enzyme production fermentors. Solid culture broths and whole Koji of *T. reesei* were also used as cellulase sources (15).

The pasteurized feed to the SSF is precooled in a heat exchanger before entering four trains of three fermentors each. With this system one train can be shut down for sterilization while maintaining the other trains in continuous operation. Over a 24 hour residence time at 40°C, the 8% cellulose feed is continuously hydrolyzed to glucose, which is simultaneously fermented to a 3.6% ethanol beer slurry. The enzyme is added as the whole culture broth without filtration or concentration. The enzyme is also recycled although the method of separation is not given. The yeast is added either as a cake or recycled as a cream. *Saccharomyces cerevisiae* (American Type Culture Collection No. 4132) and *Candida brassicae* (Institute for Fermentation, Osaka, Japan No. 1664) gave the highest ethanol yields and are the preferred strains of yeast (16).

The beer slurry is neutralized before distillation to allow most of the recovery equipment to be constructed of carbon steel. The ethanol is then concentrated to about 25% using a steam slurry stripper designed by Raphel Katzen Associates. Further concentration can then take place by normal rectification to the desired product concentration. Stillage from distillation is evaporated to a 50% solid animal feed by-product. The unconverted cellulose and lignin from SSF is the basic fuel source for the process.

b) Process Evaluation

A comparison of conversion for simultaneous saccharification and fermentation (SSF) and for saccharification only is shown in Figure IV-8. By removing glucose and cellobiose inhibition of cellulose hydrolysis, SSF increases the cellulose conversion yield approximately 25 to 40% for 24 and 48 hours, respectively, relative to saccharification only. It is not clear why only about 60% conversion is achieved for the case shown for SSF in 48 hours in Figure 8 instead of the 90% conversion assumed for design. At 10% to 20% v/v enzyme inoculation is claimed to produce 90% conversion. The specific ethanol yield per unit enzyme, however, decreases with increasing enzyme concentration (16).

Ethanol inhibition of cellulase activity was considered slightly less than by glucose based on the same weight percent (15). Since the ethanol produced is at most about 50% weight of the starting glucose, the ethanol inhibition should only be about half of the glucose inhibition of cellulase. However, the decrease in cellobiose, which is a much stronger inhibitor of cellulase than glucose, reduces the product inhibition in SSF relative to saccharification alone. Addition of up to 5% ethanol to a simple saccharification did not affect reducing sugar production implying this level of ethanol can be tolerated in an SSF (16).

The SSF design reduces the equipment cost by combining the hydrolysis and fermentation vessels into one unit. SSF also eliminates the need for separation of glucose from the hydrolysis residue. In addition, the need for strict asepsis with SSF is reduced by the ethanol and anaerobic environment

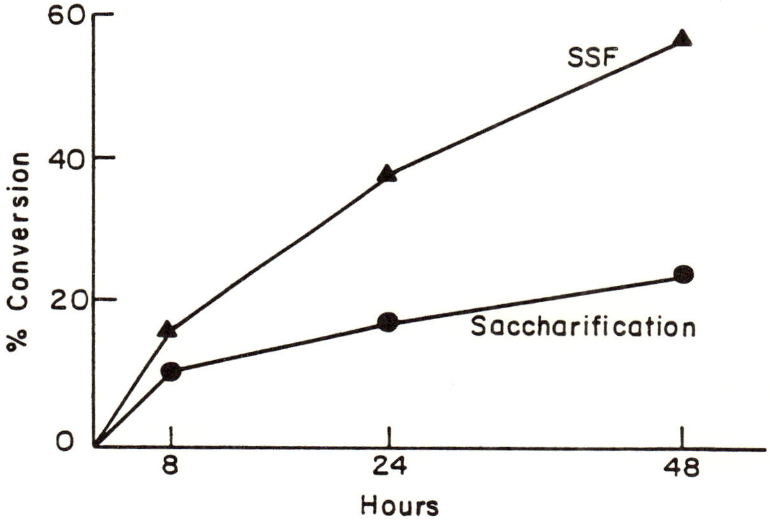

Gulf/Arkansas Comparison of Simultaneous Saccharification/Fermentation (SSF) and Saccharification

Figure IV-8.

created by the growing yeast and by the decrease in transfer operations. However, a difficulty with the SSF system is that its optimum temperature of 40°C is a compromise between the 45° to 50°C optimum for cellulase system and the 30° to 35°C optimum for most yeasts. Nevertheless, the sugar concentration in the SSF is very low and indicates hydrolysis and not fermentation is the rate limiting step.

Raphael Katzen Associates has done a technical and economic evaluation of the Gulf/Arkansas process and concluded that technical feasibility was demonstrated by Gulf. However, a critical assumption, which has raised serious doubts about the Gulf/Arkansas process is the 90% conversion of cellulose, which is assumed to be 57% of the feedstock concentration (6).

2. Mixed Culture Hydrolyzing and Fermenting Organisms Processes (MIT, GE/CRD processes)

a) Process Description

The system for simultaneous hydrolysis and fermentation can be simplified further relative to the Gulf/Arkansas Process by using a mixed culture capable of both hydrolyzing cellulose to sugars and fermenting the sugars to ethanol and side products. This process eliminates the need for a separate enzyme production vessel since the cellulolytic enzyme is produced and used in the same vessel. Fig. IV-9 shows the mode of fraction of a mixed culture system using thermophilic, anaerobic bacteria. *Clostridium thermocellum* can hydrolyze cellulose to glucose and cellobiose can hydrolyze hemicellulose to xylose and xylobiose, and can ferment glucose and cellobiose to ethanol, acetic acid and lactic acid. *Clostridium thermosaccharolyticum* does not have cellulolytic powers, but it can ferment glucose, cellobiose, xylose and xylobiose to ethanol, acetic acid and lactic acid. Therefore for a feed stock with the hemicellulose removed, for example, by acid pretreatment

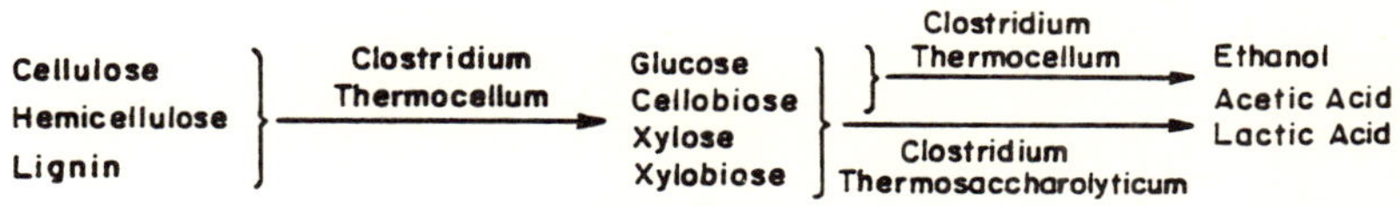

Mixed Culture Hydrolysis and Fermentation
Used by M.I.T. and GE/CRD

Figure IV-9.

or if hemicellulose conversion is not important, a single organism, C. thermocellum, can convert a cellulose feed directly to ethanol (17).

However, the two independently developed processes which use C. thermocellum also use C. thermosaccharolyticum in a mixed culture. The two conceptually similar processes are being developed by M.I.T. and General Electric Corporate Research and Development (GE/CRD). They differ mainly in the cellulose feed stock, pretreatment, possibly fermentor reactor configuration, organism strains, and recovery.

The process flow sheets for the M.I.T. and GE/CRD processes are shown in Figures IV-10 and IV-11, respectively. The M.I.T. process design and cost estimates given here were developed by Battelle, Columbus, supposedly based on experimental data from M.I.T. (18). However, there is considerable discrepancy between the laboratory results reported by M.I.T. and the design bases used by Battelle as shown in Table IV-1. The preliminary process design and cost estimate for the GE/CRD process were developed by GE/CRD. (19).

The substrates for the M.I.T. work have thus far been Solka Floc and corn stover with their process design based on corn stover; the GE/CRD process is based on poplar wood as the substrate. The corn stover is pretreated by milling for size reduction and is assumed to have a fermentable polysaccharide context of 70% on a dry wt basis. The poplar wood in the form of moist wood chips is contacted with gaseous sulfur dioxide (1% by weight of wood) under steam pressure of about 300 psi for 10 to 15 minutes.

The mixed culture fermentation with C. thermocellum and C thermosaccharolyticum is carried out at 60°C at pH 7 under strictly anaerobic conditions. The cellulase system of C. thermocellum is as effective in hydrolyzing hemicellulose to xylose as it is in hydrolyzing cellulose, but it can not utilize the xylose. C. thermosaccharolyticum is added to catabolize

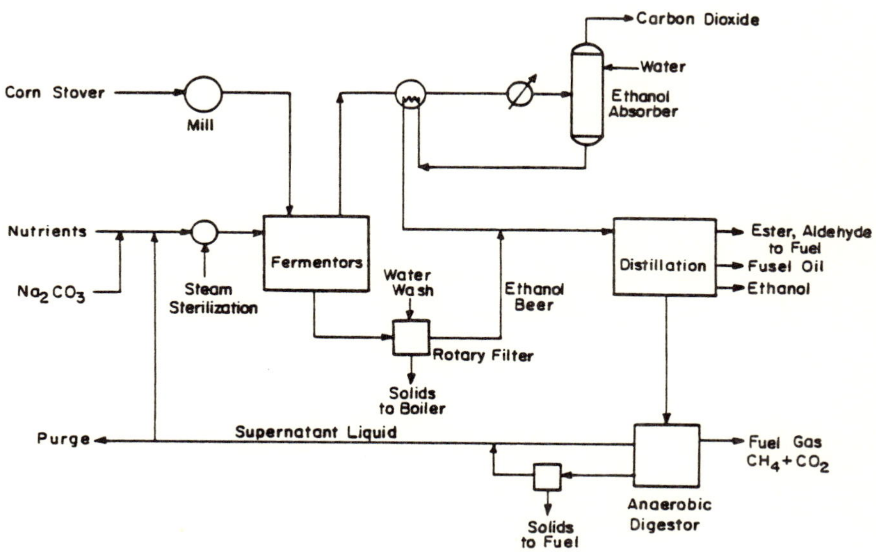

Battelle Design of M.I.T. Process

Figure IV-10.

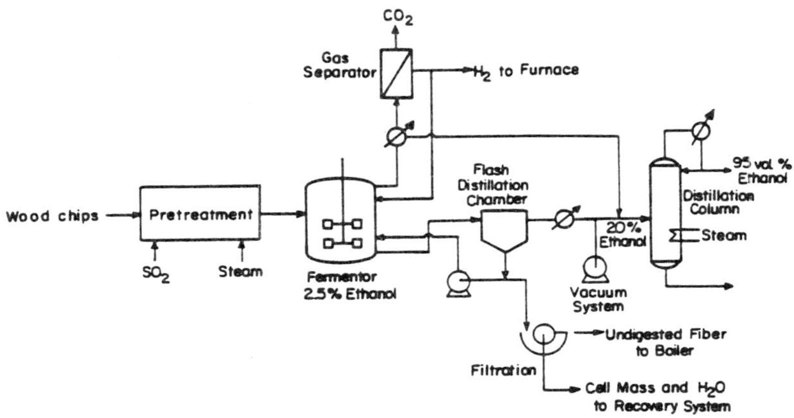

General Electric Corporate Research and Development (GE/CRD) Process

Figure IV-11.

Table IV-1

M.I. T. Yields for Mixed Cultures of *C. thermocellum* and *C. thermosaccharolyticum*

	M.I.T. Laboratory Results		Battelle Design Basis
Substrate	Solka Floc	Corn Stover (Unpretreated)	Corn Stover
Feed Method	Fed-batch	Fed-batch	Continuous (dilution rate = 0.03 hr^{-1})
Substrate Concentration (g/l)	70	100	245
Ethanol (g/l)	29	9.6	45
Acetic Acid (g/l)	6.8	6.4	22
Lactic Acid (g/l)	3.2	0.0	30
Reducing Sugars (g/l)	0.0	0.0	
Cells (g/l)			7.7
Residue (g/l)			98
Utilization of Substrate (%)	93	37	
Ethanol Productivity (g EtOH/L/hr)			1.37

xylose and is environmentally and biologically compatible with *C. thermocellum* in a stable mixed culture. At both M.I.T. and GE/CRD both organisms have undergone a program of mutation, selection, and adaptation to increase their ethanol tolerance and to alter their product distribution to produce more ethanol and less acetic and lactic acids.

The product yields and distribution achieved at M.I.T. for Solka Floc and unpretreated corn stover with current organism strains for fed-batch cultures and the assumptions for the continuous culture design case are given in Table IV-1 (11,14). The 45% conversion of Solka Floc to ethanol is encouraging, but the ethanol to acetic acid ratio is much less favorable with corn stover. M.I.T. is hypothesing that the poor product ratio for corn stover is due to a water soluble component of corn stover. Corn stover extracted with 1% NaOH resulted in a substrate conversion rate very similar to that of Solka Floc and a 3.7 to 1 ratio of ethanol to acetic acid.

In addition to the use of feed batch cultures M.I.T. has tried packed bed and expanded or fluidized bed bioreactors to achieve high solids loading. (11). The packed bed of substrate was constrained at the upper bed surface and resulted in an excessive pressure drop after time. On the other hand, mixed culture fermentations with an expanded bed with approximately 60-80 g/L biomass concentration was similar in fermentation behavior to the fed-batch cellulose fermentors.

In the Battelle design of the M.I.T. process, the solid residue from the fermentation is separated from the ethanol beer in rotary filters. The residue then is utilized as fuel for steam production after ethnaol is washed off with warm water. Ethanol is also recovered from the carbon dioxide gas after cooling in a water absorber. The ethanol is recovered further in a distillation system consisting of a stripper/rectifier, a dehydration tower, a hydrocarbon stripper, a fusel oil washer, and several heat exchangers. This distillation system is based on a Katzen design which conserves energy (20). The stillage from distillation is sent to anaerobic digestion for fuel gas production. The solids from the digestor become fuel for the steam plant, which is included in the design of off-sites facilities.

In the GE/CRD process the same mixed culture of C. thermocellum and C. thermosaccharolyticum but with different strains were used compared to the M.I.T. process. In the GE/CRD process cost estimate the following assumptions are made. The recovered fibers from pretreatment are 90% of the charge. Fifty percent of the recovered fibers are fermentable sugars of which 90% is fermented. Forty percent of the fermented sugars becomes ethanol. The ethanol concentration in the stirred tank fermentor is assumed to be 2.5%. However, the basis for the ethanol yield assumption in the GE/CRD process is not clear since none of the reported laboratory data showed ethanol concentrations near this high.

The GE/CRD process for ethanol recovery utilizes a flash ferm process (see section under "Rapid Ethanol Fermentation") in which fermentor broth is continuously pumped to a vacuum flash pot from which 20% ethanol is sent to a distillation system to produce 95% ethanol. From the flash pot the cell mass and undigested fibers are separated in a rotary filter. The cells are recycled to the fermentor to maintain high cell density, and the fibers are sent as fuel to the pretreatment steam boiler.

b) Process Evaluation

Both the M.I.T. and GE/CRD processes appear technically feasible. The concept of combining three separate operations, enzyme production, hydrolysis, and ethanol fermentation, into one reactor is certainly appealing from a processing standpoint. The probability of contamination should be reduced with the fewer number of transfer operations, especially with thermophillic anaerobic organisms. Oxygen transfer problems with aerobic cellulase production is also eliminated. The high fermentation temperature of 60°C has advantages in decreasing the vacuum requirement for ethanol distillation from the flash pot in the GE/CRD process. The continuous withdrawal of ethanol from the fermentor via the flash pot reduces the ethanol inhibition problem. Therefore, ethanol tolerance should not be as important a consideration in selection or genetic improvement of organisms for the GE/CRD process.

Problems with the M.I.T. and GE/CRD processes include maintaining low enough oxygen tension for obligate anaerobic bacteria, attaining high solids feed concentration, producing less by-products and obtaining higher ethanol yields. Much of the improvements in these areas have been accomplished and future improvements can still be expected from mutation, selection, and adaptation. Alternatively, other organisms can be considered. For example, the University of Wisconsin and M.I.T. have preliminary indications that *Chlostridium thermohydrosulfuricum* can produce high yields of ethanol from both pentoses and hexoses. A second mixed culture of this organism, with *C. thermocellum* is being studies by M.I.T.

As with the Berkeley Process, no provision was made in the M.I.T Process for large scale storage of corn stover. There may be problems of degradation unless there is protection from the weather. Also no means were provided for sterilization of the corn stover. The Berkeley process acid pretreatment simultaneously accomplishes sterilization.

C. Separate and Combined Hydrolysis and Fermentation Processes (Penn/GE Process)

1. Process Description

The University of Pennsylvania, the General Electric Co., Re-entry and Environmental Systems Div. (RESD); and Hahnemann Medical School are jointly developing an integrated process, the Penn/GE process, for the total conversion of cellulosic biomass to liquid fuels and valuable by-products. As shown in Figure IV-12, this process consists of 1) a separate hydrolysis for producing 20% sugar solution followed by fermentation to ethanol with conventional technology, 2) a simultaneous hydrolysis and extractive fermentation to produce butanol and acetone, 3) a simultaneous hydrolysis and vacuum fermentation to produce ethanol. The three sub-processes within the overall process are similar to processes described in previous sections and have similar advantages. Therefore, only major differences in advantages will be pointed out.

Refuse derived fuel (RDF) from municipal solid waste processing is the cellulosic feedstock immediately being considered. Wood chips from fast growing poplar trees are considered the long term feedstock. Pretreatment for all three sub-processes consists of lignin extraction with a hot aqueous organic solvent mixture. Butanol has been tried and ethanol will be tried next as the organic solvent. The delignified cellulose solids from pretreatment are sent to the enzyme hydrolysis vessel. The aqueous phase contains partially degraded hemicellulose from which relatively pure linear xylans can be recovered by precipitation following acidification. The xylan can then be hydrolyzed by xylanase derived from *Thermomonospora* to produce xylose, a valuable by-product. The remaining hemicellulose degradation products in the aqueous stream are fermented by *Clostridium acetobutylicum* in simultaneous hydrolysis and extractive fermentation to yield butanol and some acetone. The organic phase assuming butanol as the solvent contains the lignin, which can be recovered as a polymer grade. The remaining lignin and butanol forms a pumpable slurry with high heating value.

Enzyme hydrolysis of pretreated cellulose solids is supposed to produce greater than 20% glucose syrups. The enzyme system for the hydrolysis is

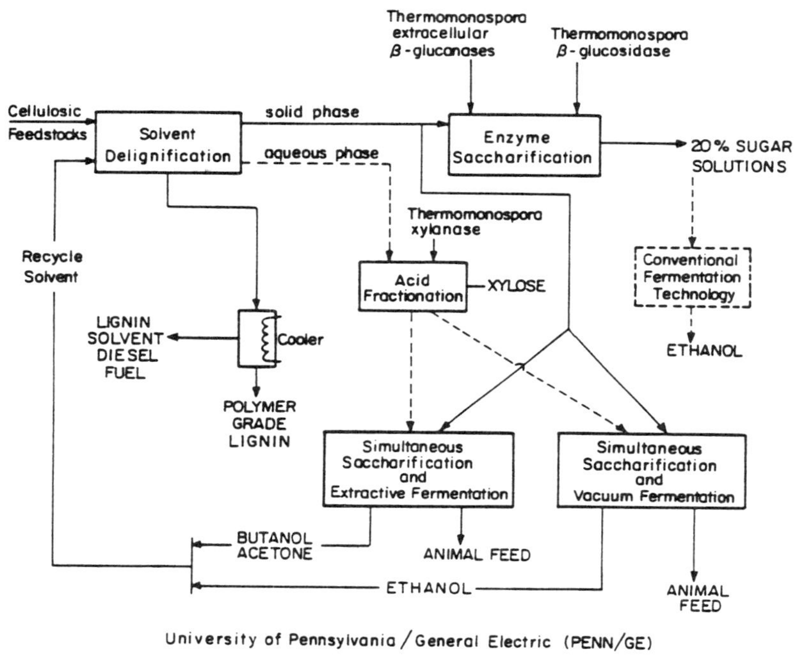
Thermomonospora extracellular β-glucanases
Thermomonospora β-glucosidase
Cellulosic Feedstocks
Solvent Delignification
solid phase
aqueous phase
Enzyme Saccharification
20% SUGAR SOLUTIONS
Thermomonospora xylanase
Recycle Solvent
Acid Fractionation
XYLOSE
Conventional Fermentation Technology
ETHANOL
LIGNIN SOLVENT DIESEL FUEL
Cooler
POLYMER GRADE LIGNIN
Simultaneous Saccharification and Extractive Fermentation
Simultaneous Saccharification and Vacuum Fermentation
BUTANOL ACETONE
ANIMAL FEED
ETHANOL
ANIMAL FEED
University of Pennsylvania/General Electric (PENN/GE)

Figure IV-12.

is derived from Thermomonospora which produces maximum activity enzyme in less than 24 hours when cultured on cellulose. The enzyme system consist of extracellular β-glucanases, which hydrolyze cellulose to cellobiose, and cell associated β-glucosidase, which hydrolyzes cellobiose to glucose. Cellobiose is very inhibitory to the β-glucanases. Glucose inhibits β-glucosidase relatively less.

The reactor configuration currently being tested for the enzyme hydrolysis process is shown in Figure IV-13. It consists of a stirred tank reactor for cellulose hydrolysis followed by an immobilized β-glucosidase fixed bed reactor, from which most of the solution is recycled to the stirred tank for high conversion yields. Use of an immobilized β-glucosidase reactor has also been studied by the University of Connecticut (21), Natick (22), and Berkeley (23).

A second path for cellulose to ethanol conversion employs simultaneous hydrolysis and vacuum fermentation. The high temperature stability of the Thermomonospora β-glucanases allows them to hydrolyze cellulose to cellobiose in conjunction with hydrolysis and fermentation of cellulose to ethanol by an anaerobic thermophilic bacterium, C. thermocellum. The maximum ethanol yield achieved, thusfar, with C. thermocellum, strain 651, is 1.0 g/L with about 90% sugar utilization, but with no xylose utilization. A noncellulolytic strain of Clostridium was able to grow on xylan at 60°C in the presence of xylanase from Thermomonospora. The vacuum used in this process is not applied to the fermentor directly, but rather on a side arm such that the CO_2 and H_2 can be removed at atmospheric pressure and not add to the compressor load. The H_2 concentration in the CO_2 and H_2 mix is high enough to provide process heat when burned.

A third path for the cellulose and unutilized degraded hemicellulose

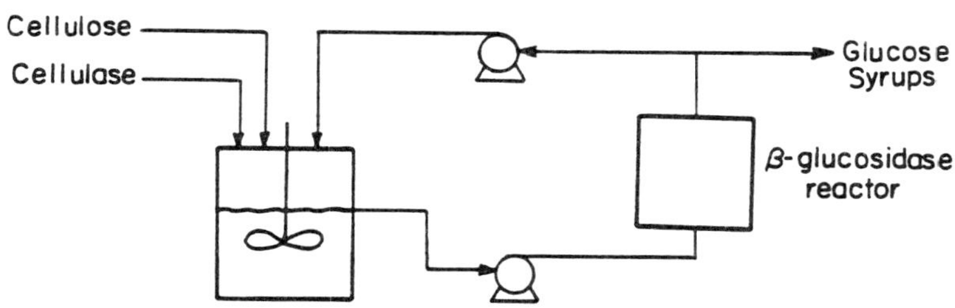

Penn/GE Hydrolysis with Separate β-glucosidase Reactor

Figure IV-13.

from delignification is conversion to butanol and acetone in a combined hydrolysis and extractive fermentation. The process utilizes enzyme from *Thermomonospora* for cellulose and hemicellulose hydrolysis in combination with *Clostridium acetobutylicum* for fermentation of glucose and xylose to butanol and acetone. The butanol at 1.5 to 2.0% is inhibitory to the organism and so is removed by continuous extraction. The extractant absorbs the butanol either directly or through a membrane and the mixture is then separated by flash distillation. After cooling, the extractant returns to the fermentor and the butanol goes back into the delignification process. Dibutyl-phtalate appears to be a suitable extractant for this process.

2. *Process Evaluation*

The Penn/GE process offers total biomass utilization and great flexibility in producing a wide distribution of products, namely, high concentration sugars ethanol, butanol, acetone, lignin-butanol fuel, and animal feed. The process also makes use of high potential innovations in integrating hydrolysis, fermentation, and recovery steps. However, the current emphasis appears to be on development of processes which will most immediately increase ethanol productivity in the United States in the near future. In particular, the solvent pretreatment and high glucose concentration processes are being emphasized and appear to be the most developed sections of the total process. For improving enzyme hydrolysis of cellulose to high concentration glucose it is important to increase the activity of the *Thermomonospora* cellulase, which is significantly lower than that from *Trichoderma reesei* (24), and to increase the stability of β-glucosidase.

An interesting innovation to the enzyme hydrolysis process being considered is use of a three reactor system consisting of 1) cellulase hydrolysis of cellulose to cellobiose in a stirred tank reactor, 2) combined cellobiose hydrolysis to glucose and glucose fermentation to ethanol by immobilized β-glucosidase and immobilized yeast on the same column, and 3) an ethanol extraction system.

The overall process is still essentially conceptual. Very little data has been reported for operation of the sub-processes as conceptualized. Most of the processes have not been in operation or are in the beginning stages of operation at the bench scale. Nevertheless, it was reported at the Third Annual Biomass Energy Systems Conference that the integrated process is expected to be demonstrated on the bench scale within one year (25).

D. Comparison of Processes and Conclusions

Based on the information presently available, it appears that Natick and Gulf/Arkansas have the most fully developed processes, which are the most likely to be commercialized in the near future. The major advantages of the Gulf/Arkansas process are the increased hydrolysis yield and the elimination of a separate fermentation vessel from utilizing simultaneous hydrolysis and fermentation. The major advantage of the Natick Process is its high enzyme productivity. If its enzyme can be made resistant to end product inhibition from glucose and cellobiose during hydrolysis, the Gulf/Arkansas advantages are greatly reduced. The best process would combine the Gulf/Arkansas processing steps with the hypercellulase producing organisms of the Natick process. Additional modifications of the process could come from substituting the pretreatment methods of other processes as discussed in the pretreatment section. Furthermore, modifications may be necessary to deal with waste cellulose feeds contaminated with toxic chemicals and possible inhibitory factors to fermentation formed during enzyme production. In addition to the enzymatic processes described in this report, there are proprietary processes commercially available at present.

Some of these processes are available from the companies listed under the section on "Rapid Ethanol Fermentation."

The M.I.T., GE/CRD and PENN/GE processes offer important innovations which require more development time but have high potential for the future. The M.I.T. and GE/CRD processes simplifies conversion of pretreated cellulose to ethanol to a single reactor. Improvements are still needed in ethanol yield and selectivity. The PENN/GE process integrates a number of options for total biomass uitlization and offers flexibility in the distribution of products, but requires a great deal of development work with both organisms and processing.

References

1. L. Spano, T. Tassinari, D.D.Y Ryu, A. Allen, and M. Mandels, "Enzymatic Hydrolysis of Cellulose to Fermentable Sugar for Production of Ethanol," Presented at the 29th Canadian Chemical Engineering Conference at Sarnin, Ontario, October 1979.

2. C.R. Wilke, R.D. Yang, A.F. Sciamanna, and R.P. Freitas, "Raw Materials Evaluation and Process Development Studies for Conversion of Biomass to Sugars and Ethanol," Proceedings of the Second Annual Symposium on Fuels from Biomass, Dept. of Energy meeting, Troy, New York, June 20-22, 1978, Volume 1.

2. T.K. Ghose and P. Ghosh, "Bioconversion of Cellulosic Substances," J. Appl. Chem. Biotechnol. 28, 309-320 (1978).

3. T.K. Ghose and J.A. Kostick, "A Model for Continuous Enzymatic Saccharification of Cellulose with Simultaneous Removal of Glucose Syrup," Biotechnol. Bioeng. 12, 921-946 (1970).

5. B.S. Montencourt and D.E. Eveleigh, "High Cellulolytic Mutants and Their Role in Saccharification," Proceedings from the Second Annual Symposium on Fuels from Biomass, June 20-22, 1978, Troy, New York, Vol. 2.

6. A.E. Humphrey and E.J. Nolan, Report to the Office of Technology Assessment, Biological Production of Liquid Fuels and Chemical Feedstocks, Govt. Printing Office, #052-003-00706.

7. I. Karube, S. Tanoka, T. Shirai, and S. Suzuki, "Hydrolysis of Cellulose in a Cellulose-Bead Fluidized Bed Reactor," Biotechnol. Bioeng. 19, 1183-1191 (1977).

8. W.W. Pitt, Jr., C.W. Hancher, and H.W. Hsu, "The Tapered Fluidized Bed Bioreactor--An Improved Device for Continuous Cultivation," AIChE Symp. Series, Vol. 74, No. 181, 119-123 (1978).

9. D.J. O'Neil, et al., "Design, Fabrication and Operation of a Biomass Fermentation Facility," First Quarterly Report, Georgia Inst. of Tech. Prepared for U.S. Dept. of Energy, contract no. ET-78-C-01-3060.

10. J.E. Sanderson, D.U.G. Martinez, J.J. Dillon, and G.S. George, and D.L. Wise, "Liquid Fuel Production from Biomass," Third Annual Biomass Energy Systems Conference Proceedings, June 5-7, 1979, Solar Energy Research Institute, Golden, Colorado.

11. D.I.C. Wang, C.L. Cooney, A.L. Demain, R.F. Gomez, and A.J. Sinskey, "Degradation of Cellulosic Biomass and its Subsequent Utilization for the Production of Chemical Feedstocks," September 1979.

12. N. Toyama, "Feasibility of Sugar Production from Agricultural and Urban Cellulosic Wastes with Trichoderma viride Cellulase," Biotechnol. Bioeng. Symp. No. 6, 207-219 (1976).

13. G.H. Emert, and R. Katzen, "Chemicals from Biomass by Improved Enzyme Technology," presented at ACS/CSJ Joint Chemical Congress, April 1-6, 1979, Honolulu, Hawaii.

14. G. H. Emert, R. Katzen, R.E. Fredrickson, and K.F. Kaupisch, "Economic Update of the Bioconversion of Cellulose to Ethanol," presented at American Institute of Chemical Engineers, 72nd Annual Meeting, Nov. 25-29, 1979, San Francisco, Ca.

15. M. Takagi, S. Abe, S. Suzuki, G.H. Emert, and N. Kato, "A Method for Production of Alcohol Directly from Cellulose Using Cellulase and Yeast," Proc. Bioconversion Symp., IIT Delhi, 551-571 (1977).

16. P.J. Blotkamp, M. Takagi, M.S. Pemberton, G.H. Emert, "Enzymatic Hydrolysis of Cellulose and Simultaneous Fermentation to Alcohol," AIChE Symp. Series, Vol. 74, No. 181, 85-90 (1978).

17. C.L. Cooney, D.I.C.Wang, S. Wang, J. Gordon, and M. Jiminez, "Simultaneous Cellulose Hydrolysis and Ethanol Production by a Cellulolytic Anaerobic Bacterium," Biotechnol. Bioeng. Symp. No. 8, 103-114 (1978).

18. D.M. Jenkins, and T.S. Reddy, "Economic Evaluation of the M.I.T. Process for Manufacture of Ethanol," Report to U.S. Department of Energy from Battelle, Columbus Laboratories, Columbus, O., June 28, 1979.

19. R. Brooks, T. Su, M. Brennan, and J. Frick, "Bioconversion of Plant Biomass to Ethanol," Third Annual Biomass Energy Systems Conference Proceedings, June 5-7, 1979, SERI, Golden, Colorado.

20. G.D. Moon, et al., "Grain Motor Fuel Alcohol Technical and Economic Assessment Study," Raphael Katzen Assn. report to U.S. Dept. of Energy, December 1978.

21. H.E. Klei, D.W. Sundstrom, R.W. Coughlin, K. Ziolkowski, and G. Biederman, "Improvement of Yields and Rates Using Immobilized Enzymes During the Hydrolysis of Cellulose to Glucose," Third Annual Biomass Energy Systems Conference Proceedings, June 5-7, 1979, Solar Energy Research Inst., Golden, Colorado.

22. F. Bissett, and D. Sternberg, Appl. and Env. Microbiol., 35, 750 (1978)

23. S.H. Isaacs and C.R.Wilke, "Utilization of Immobilized β-Glucosidase in the Enzymatic Hydrolysis of Cellulose," Lawrence Berkeley Laboratory, Univ. of California, Berkeley, LBL-7857, 1978.

24. E.K. Pye, and A.E. Humphrey, "Biological Production of Liquid Fuels from Biomass," Univ. of Pennsylvania Interim Report, (June 1 - August 31, 1979) prepared for U.S.Dept. of Energy.

25. E.K Pye, and A.E. Humphrey, "Production of Liquid Fuels from Cellulosic Biomass, Third Annual Biomass Energy Systems Conference Proceedings, June 5-7, 1979, SERI, Golden, Colorado.

V. High Productivity Ethanol Fermentation

INTRODUCTION

A. Conventional Process Limitations

The total world industrial ethanol production is approximately 800 million gallons per year (1). Three quarters of the production is by batch fermentation and, with only rare exception, the same batch techniques employed at the turn of the century are still used (2).

Batch fermentation is a slow process. The fermentation vessel is first cleaned and prepared. Typically, a 10-15 w/v% solution of pasteurized sugar supplemented with yeast nutrients is added to the vessel and the vessel is inoculated with a rapidly growing culture of yeast from a seed tank. The course of a fermentation using a high productivity yeast strain is plotted in Figure V-1 (3). The inoculation yeast continue to multiply in the fermentor simultaneously producing alcohol. The rate of alochol production is initially quite low, but as the number of yeast cells increases, the overall rate of fermentation increases. A maximum in ethanol productivity is reached after 10 hours. Ethanol production then continues at a decreasing rate until at 14 hours, 94% of the sugar is utilized, and a final ethanol content of 4.6 g ethanol/L is achieved. The fermentor beer is then emptied into a holding tank to be fed to the continuous distillation system. The fermentor is cleaned and prepared for another fermentation cycle.

The overall productivity for the batch fermentation process is typically only 1.8 to 2.5 grams of ethanol produced per liter of fermentor volume per hours, (4,5) and eighteen 100,000 gallon batch fermentor are required for a 25 million gallon per year alcohol fuel plant. Designs for such a plant are available in the literature (4,6,7,8). The capital cost of the fermentation section is estimated at approximately 17 million dollars (7,9). Finance and

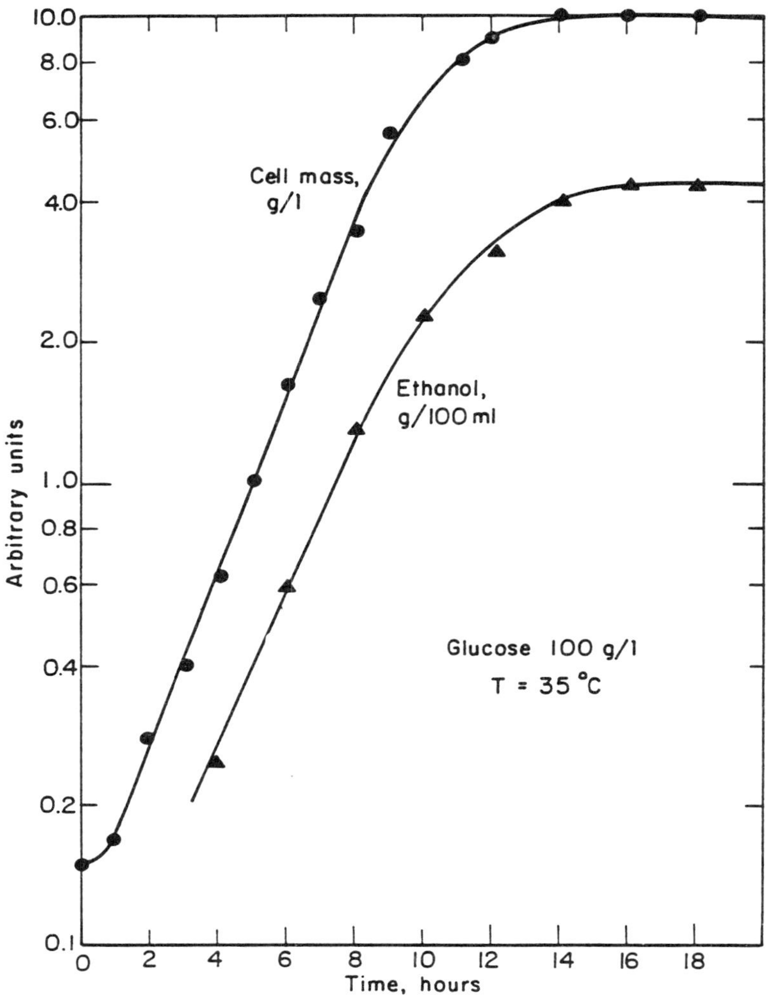

Figure V-1. Batch Fermentation with Saccharomyces cerevisiae, var. anamensis.

depreciation cost in support of such a plant contribute 5¢/L to the cost of alcohol (7). The continual start up/shut down nature of the batch process makes it difficult to automate and high labor costs also result.

The need for new processes is apparent. Fortunately, great progress has been made with as much new fermentation research conducted over the past fifteen years at in the previous fifty (4).

B. The Theory of Fermentation with Yeast

An understanding of the basic fermentation reactions and their regulation is helpful in the evaluation of alternative fermentation processes.

1. Yeast Metabolic Pathways

A schematic representation of the anaerobic and aerobic yeast metabolic pathways is presented in Figure V-2 (10).Under anaerobic conditions, glucose is fermented to ethanol and carbon dioxide by glycolysis. The overall reaction to liberate energy for biosynthesis, (Equation 1) produces 2 moles of ethanol and carbon dioxide for every mole of glucose consumed.

Theoretical:

$$C_6H_{12}O_6 \rightarrow 2C_2H_5OH + 2CO_2 + \text{Energy (Stored as ATP)} \qquad (1)$$

On a weight basis, every gram of glucose can theoretically yield 0.51 grams of ethanol. In practice though, actual ethanol yield are about 90% of the theoretical--a portion of the glucose carbon source being used for synthesis of new cell mass (Equation 2).

Observed:

$$1g\ (C_6H_{12}O_6) \rightarrow .460\ g\ (C_2H_5OH) + .460\ g\ (CO_2 + 0.10\ g\ (\text{new cells}) \qquad (2)$$

Under aerobic conditions, glucose is converted completely to carbon dioxide and new cell mass, with no ethanol being formed.

Simplified Chart of Anaerobic and Aerobic Catabolism of *Saccharomyces cerevisiae*

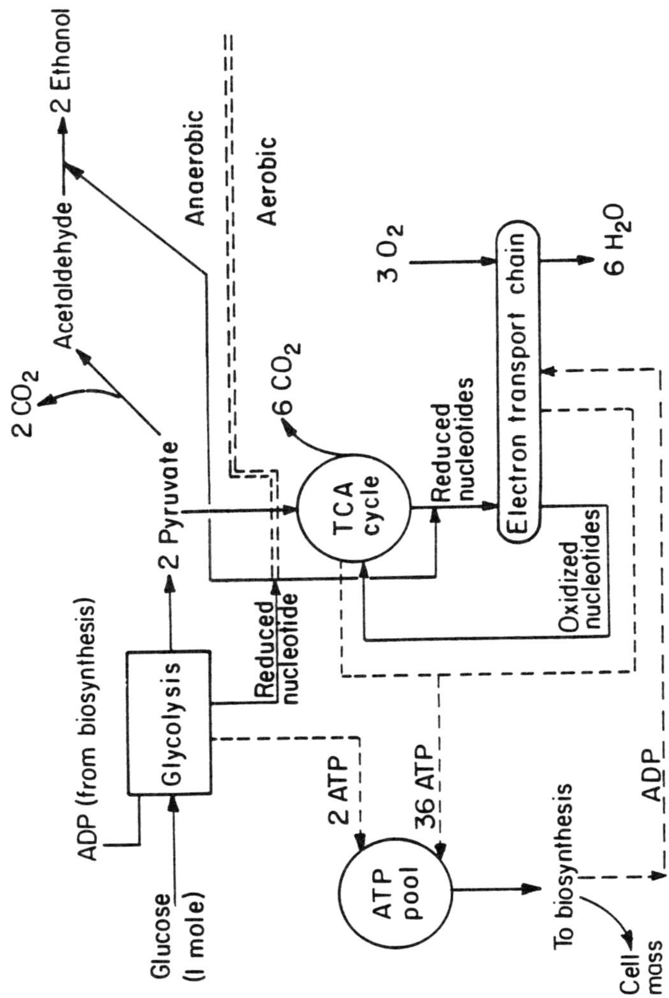

Figure V-2.

2. Effect of Glucose Concentration

Glucose is the primary reactant in the yeast metabolism and at very low glucose concentrations (less than 10 mg/L) the rate of glucose consumption increases roughly linearly with glucose concentration (11). A high concentrations (above 150 g/L) glucose inhibits enzymes in both the fermentative and oxidative pathways (12). At intermediate glucose concentration (3 to 100 g/L) catabolite repression of the oxidative pathways takes place, permitting production of ethanol even in the presence of large amounts of oxygen (13,14,15).

Under anaerobic growth conditions, caused either by catabolite repression or by oxygen restriction, and at low to moderate glucose levels, the rate of ethanol production is given roughly by a Monod type relationship.

$$V = V_{max} \frac{S}{K_s + S} \tag{3}$$

where;

V is the ethanol production rate in (g ethanol/g cells - hr).

S is glucose substrate concentration (g/L). The constant factor.

K_s has a very low value on the order of 0.025 g/L, so that under anaerobic conditions the rate of ethanol productivity per cell is essentially at its maximum value for any sugar concentrations above 0.2 g/L, and less than 150 g/L (where inhibition becomes important).

3. Effect of Oxygen

It is important to avoid aerobic metabolism which utilizes glucose feed but produces no ethanol. At high glucose concentrations, catabolite repression assures fermentative metabolism. At low sugar concentrations, the oxygen concentration must be restricted. Oxygen ordinarily should not be totally eliminated, however, as oxygen is a necessary building block for the biosynthesis of polyunsaturated fats and lipids required in mitochondria and plasma membranes (16,17,18,19). Trace amounts of oxygen (0.05-0.10 mm Hg oxygen tension)in the fermentor beer are adequate (3,13,20,21,22) and do not promote aerobic metabolism. As oxygen is continually consumed by the yeast during growth, it is normally necessary to sparge the fermentor with air to maintain even these trace oxygen levels.

4. Effect of Ethanol

Ethanol, the product of fermentation, is itself a yeast toxin. The tolerance of various yeasts to ethanol depends considerably on the strain chosen, but ethanol production and cell growth are generally brought to a complete halt at ethanol levels greater than 120 g/L (3,23,24). For most yeast, the effect of ethanol inhibition is negligible at low alcohol concentrations (less than 30 g/L) but increases rapidly at higher concentrations (Fig. V-3, 3A)(24,25,26,27).

5. Total Versus Specific Productivity

Thus far, we have discussed specific productivity, the rate of ethanol production per yeast cell. Specific productivity is a function of glucose concentration, oxygen concentration and ethanol concentration (as well as of many process independent variables such as temperature and pH). An industrial fermentation must maximize total productivity, which is the product of specific productivity and yeast cell concentration.

Typical cell concentrations at the end of a batch fermentation are of the order of only 5 to 10 g/L. Continuous fermentation experiments have been conducted at levels up to 120 g/L and the expected increases in total ethanol productivity have been observed (28,29). At higher cell densities, approaching the maximum packing density of 155 g/L (30), transport limitations (31) and

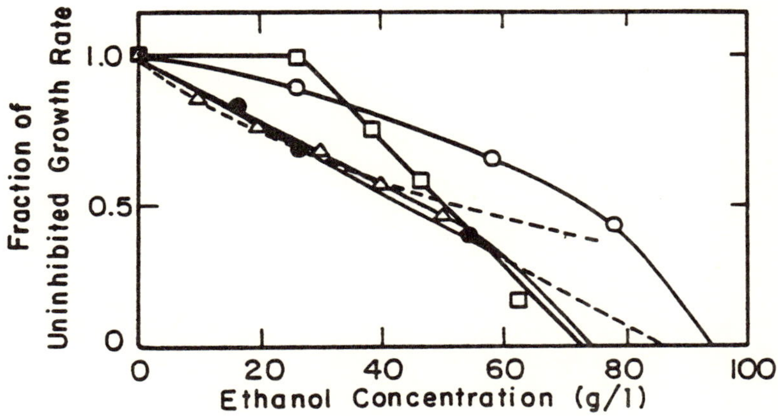

Figure V-3. Ethanol Inhibition of Çell Growth (□ Holtzberg, ○ Bazua and Wilke, --- Aiba, △ Aiba and Shoda, ● Ghose and Tyagi)

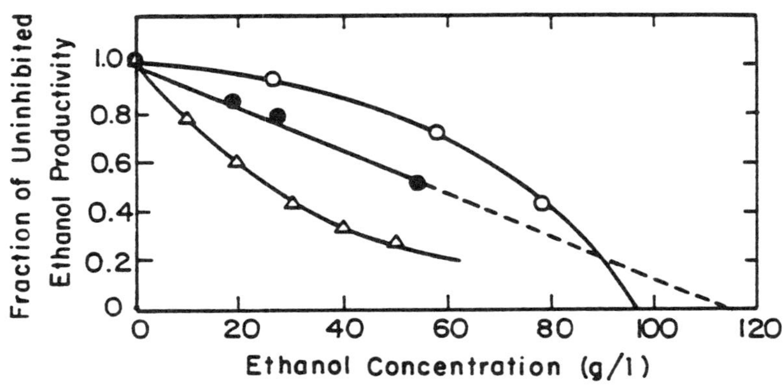

Figure V-3A. Ethanol End Product Inhibition (○ Bazua and Wilke, △ Aiba, ● Ghose and Tyagi).

possible cellular inhibition are anticipated (32,33).

C. Conditions for High Rate Fermentation

With the basic yeast fermentation reaction and its regulation by ethanol, oxygen, and glucose concentrations outlined, we can now propose those conditions which would lead to a high rate fermentation.

The specific productivity must be high. Therefore, the oxygen level should be regulated to prevent aerobic growth, but to adequately meet yeast oxygen maintenance requirements. The ethanol level in the fermentor should be maintained low, so as not to inhibit further production. Glucose concentrations should be maintained at not less than 0.2 g/L so as not to inflict a feed limitation. Finally, the concentration of yeast cells (the catalyst for the ethanol production reaction) should be maintained as high as possible without causing transport associated limitations.

HIGH RATE FERMENTATION PROCESSES

A. Criteria for Evaluation of Potential Industrial Processes

Very few of the high rate fermentation processes under development have been advanced to the point of pilot plant testing. Detailed economic evaluations of the potential processes are therefore not available and another means of comparison is required. Some simple criteria for evaluating the potential of new processes for industrial application are presented in Table V-1

Fully continuous processes offer many advantages (34, 35, 36). 1) Fermentation can be conducted at a single optimum condition (not varying with time), 2) Greater throughput of product and consistency of product quality are possible, 3) Manpower is reduced as cleaning and refilling operations are eliminated, 4) Total production is increased ad downtime between cycles is eliminated, 5) A steady demand is placed on electricity, steam and other services, thus eliminating high peak load levels, 6) Automation is simplified based on real time sampling of the final product, 7) Intermediate

Table V-1

Some Criteria for Comparing Fermentation Processes

Low Operation Cost:

1. Continuous process
2. Simple operation
3. Low Energy Input
4. Near Complete Sugar Utilization

Low Capital Cost:

5. High Productivity (small fermentor volume)
6. Mechanically simple

product storage is not required for buffering to interface with continuous feed pasteurization and product concentration units, 8) The process is more easily controlled and adjusted to meet varying requirements. For these many reasons, continuous fermentation processes are to be favored for industrial application.

Simple processes (with few steps) reduce maintenance requirements. Control is made simpler and these processes should be favored.

Energy and glucose feed costs contribute substantially to the cost of the final product. Processes which utilize little energy (for pumps, refrigeration, etc.) and which utilize the glucose feed as completely as possible (minimizing waste) are clearly advantageous.

The special advantages of high rate fermentation processes come in reduced capital and maintenance costs. High-rate processes allow the use of much smaller fermentation equipment. When the process is also mechanically simple, major capital cost savings are assured.

B. Alternative Processes

1. The Simple Continuous Stirred Tank Fermentor

The continuous stirred tank fermentor (C.S.T.R.) is a simple point of departure from standard batch process.

A C.S.T.R. fermentor is depicted schematically in Figure V-4. Pasteurized feed is pumped continuously into an agitated vessel in which yeast are actively fermenting. The sugar is largely consumed and ethanol and new cell mass are produced. Fermentor beer containing ethanol, yeast cells, and residual sugar flows continuously from an overflow port in the side of the fermentor. Air is sparged through the fermentor base to maintain the optimum oxygen tension. The composition of the beer in the fermentor is everywhere uniform and is the same as the composition of the overflow stream (37).

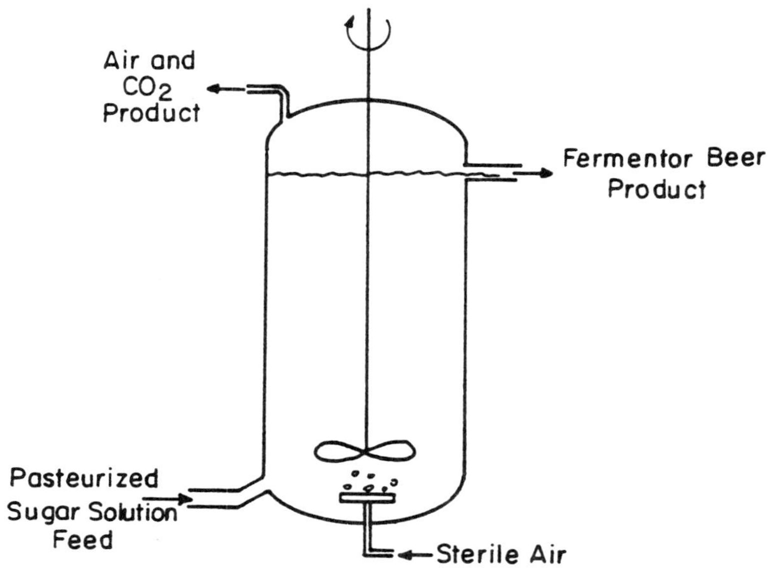

Continuous Stirred Tank Fermentor

Figure V-4.

Specific ethanol productivity in the simple CSTR fermentor is ordinarily limited by ethanol inhibition (28). To avoid very high ethanol distillation costs, the ethanol overflow product concentration and hence the concentration in the fermentor must be maintained at a relatively high level. Cysewski has found an optimum of 5 w/v% to give minimum fermentation plus distillation cost for one yeast strain (6).

The total productivity of a CSTR fermentor system is also limited by the low cell densities achieved in the fermentor. New cells are continuously born in the fermentor, but cells are also continuously washed out. A steady state is achieved when the growth and washout rates are identical. A cell density of only 10-12 g/L is typical (28). The overall productivity for a simple CSTR fermentor using a high productivity yeast is approximately 6 g ethanol/L-hr, three times the average batch productivity (6).

Considered in terms of our evaluation criteria the simple CSTR process offers many advantages over conventional batch technology. The process is continuous, involves few steps, has no unusual energy requirements and can achieve virtually complete sugar utilization. Reduced operating costs are expected. The equipment is mechanically very simple and total required fermentor volumes are only 1/3 those required for batch operation. Capital costs should, therefore be reduced. These predictions are supported by a detailed economic analysis by Cysewski which shows a 53% reduction in operating costs and a 50% reduction in capital costs for the CSTR process as compared to conventional batch fermentation (6).

2. Series CSTR Fermentors

Advantages have been demonstrated for CSTR fermentors arranged in series both at laboratory (25,34,35) and plant scale (2,39). Consider first two CSTR fermentors in series. Sugar solution is fed to the first and fermentation takes place. The residence time is adjusted so that the sugar is only partly utilized in the first fermentor. The ethanol concentration in the fermentor is thus less than for complete utilization. The overflow from the first fermentor is fed to the second, where fermentation is completed producing the final high ethanol concentration beer product. Because the first fermentor is operated at reduced ethanol concentration, ethanol inhibition is reduced and the productivity of the first fermentor is quite high. The second, lower productivity fermentor, now must convert less sugar than if it was operated alone. The result is an overall increase in productivity compared to a single vessel. Ghose and Tyagi (25) have shown the productivity of a two-stage CSTR system to be 2.3 times that of a single CSTR.

Systems with several vessels in series can also be considered.

The concept of multiple series CSTRs has been applied in the multistage perforated plate column fermentor (Fig. V-5)(90,91,92,93,94). The fermentor consists of a column divided into stages by perforated plates. A single shaft drives impellers at all stages. Each stage acts as an individual fermentation vessel. Feed added at the column head is partly fermented in the first stage and the trickles down and is successively fermented through the lower stages.

Like the simple CSTR system, simple two-stage series CSTR systems appear advantageous based on the selection criteria. It is not clear though, whether the added complications of a multiple vessel system will justify the increase in productivity and the overall reduction in fermentor volume.

The mechanical complexity of the perforated plate column fermentor is a major factor against its industrial application.

3. Continuous Cell Recycle Reactors

The use of continuous cell recycle reactors has been thoroughly investigated (6,9,24,40) and has been applied at large scale (41). A cell

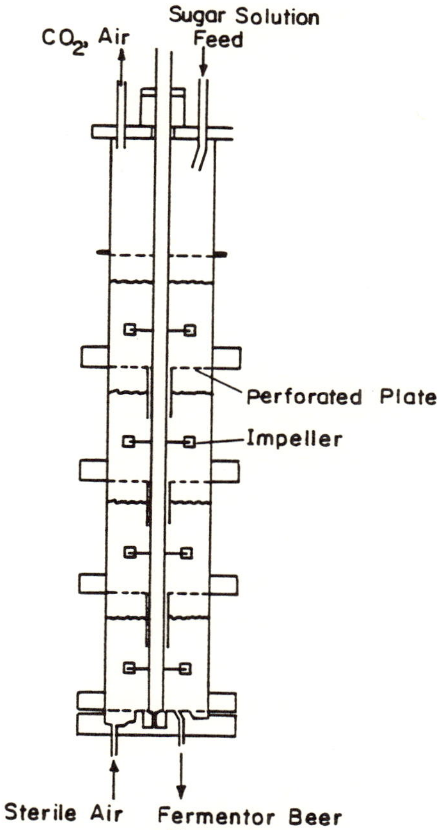

Multistage Perforated Plate Column Fermentor

Figure V-5.

recycle system is shown in Figure V-6. The system is identical to the simple CSTR fermentor, except that a centrifuge is used to separate yeast from the product overflow and return this yeast to the fermentor vessel. With continued cell growth, and cell escape prevented, the cell concentration in the fermentor becomes extremely high and total productivity is greatly increased. A small bleed of cells is required to maintain a viable culture (11,42), but cell densities as high as 83 g/L can be maintained (3,29). Higher concentration sugar feeds can be fermented, and fermentor productivities of 30 to 40 g/L-hr are possible (25,28,29,43).

This continuous process is also quite attractive. Some added complexity results from the need for a mechanical centrifuge which increases capital cost and requires considerable maintenance. Electrical energy costs are increased, and added supervision is required to monitor the centrifuge. These disadvantages have been shown to be more than offset by the great increase in productivity and reduction in equipment size (9).

Several attempts have been made to develop simplified cell recycle systems which do not require mechanical centrifuges. Simple cell settling systems have been proposed wherein the cells are thermally shocked (to temporarily halt CO_2 evolution) and allow to gravity settle from the fermentor overflow. Very large settling vessels are required, however (44, 45).

Whirlpool separators have been investigated (Fig. V-7)(4). Yeast cells are deposited in a central cone when the fermentor overflow is pumped tangentially into a verticle cylindrical vessel. Energy requirement for the Whirlpool separator are low (46).

A very simple partial recycle fermentor has been developed and tested at pilot scale (Fig. V-8)(47).The overflow is taken from a verticle pipe

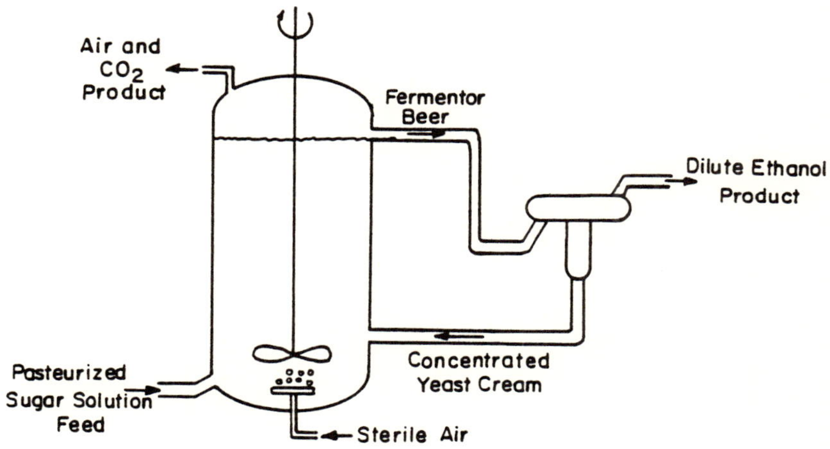

Continuous Stirred Tank Fermentor with Yeast Cell Recycle

Figure V-6.

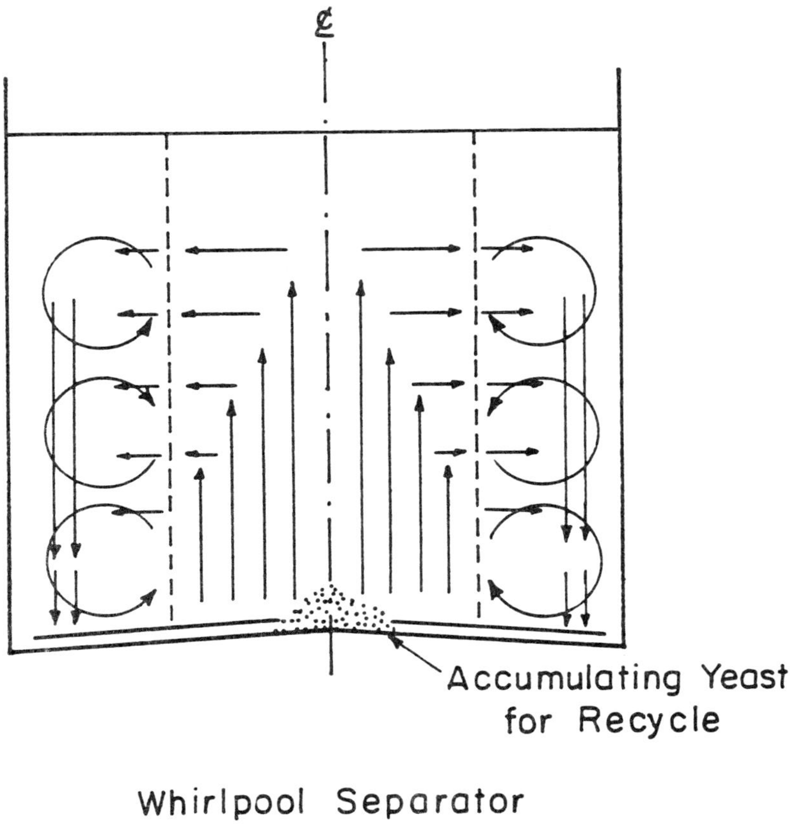

Figure V-7.

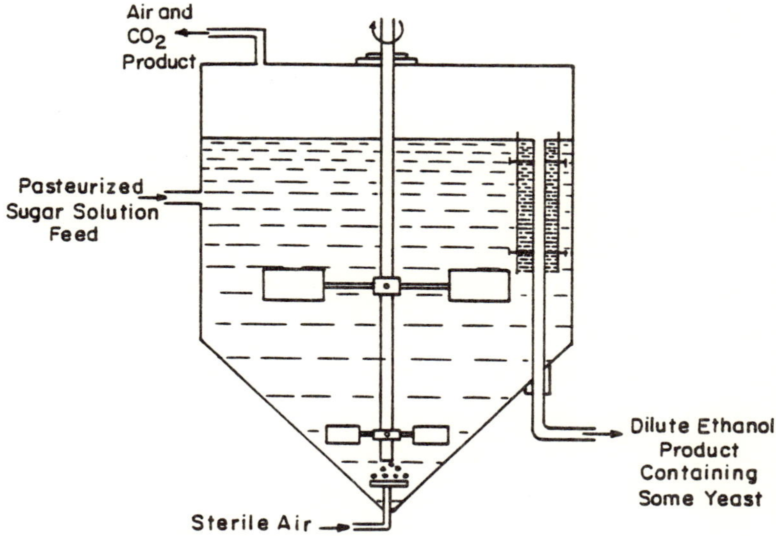

Partial Recycle Reactor

Figure V-8.

rising through the fermentor base. This pipe is jacketed by a baffled sleeve. The region between pipe and sleeve escapes agitation and yeast tend to separate from the beer rising to the overflow nozzle. The separation is far from complete, but some concentration of cells in the fermentor is achieved, and with essentially no added equipment.

4. Tower Fermentors

Tower fermentors of many designs have been tested. The APV tower system (Fig.V-9)has been operated successfully at large scale (48). The fermentor consists of a verticle cylindrical tower with a conical bottom. The tower is topped by a large diameter settling zone fitted with baffles. The overall aspect ratio is from 7:1 to 10:1 with tower diameters from 0.9 to 2 meters (49,50,51,52). Sugar solution is pumped into the base of the tower, which contains a plug of flocculent yeast (48). Fermentation proceeds progressively as the beer rises, but yeast tendsto settle back and be retained. The limitation on beer throughput is set by the requirement that yeast be effectively retained by settling against the upward flow. High cell densities of 50 to 80 g/L are achieved (48) without the requirement of an auxiliary mechanical separator. Productivities 32 to 80 times those for simple batch fermentors have been achieved with the APV tower fermentor system (2)*. Some difficulty is associated with providing the desired oxygen concentration in the fermentor as direct sparging at the base promotes turbulence and inhibits yeast settling.

*Caution must be used in comparing productivities for various fermentation schemes. Different yeasts and sugar feed solutions have been used in testing the various systems. Productivities 80 times higher than for simple batch fermentation have been achieved in the tower fermentor comparing slow growing brewer's yeasts in both cases. The productivity of 83g ethanol/liter-hour achieved by Cysewski with vacuum fermentation is only 80 times the batch productivity achieved with the same high productivity yeast--yet, this is the highest absolute fermentation rate reported.

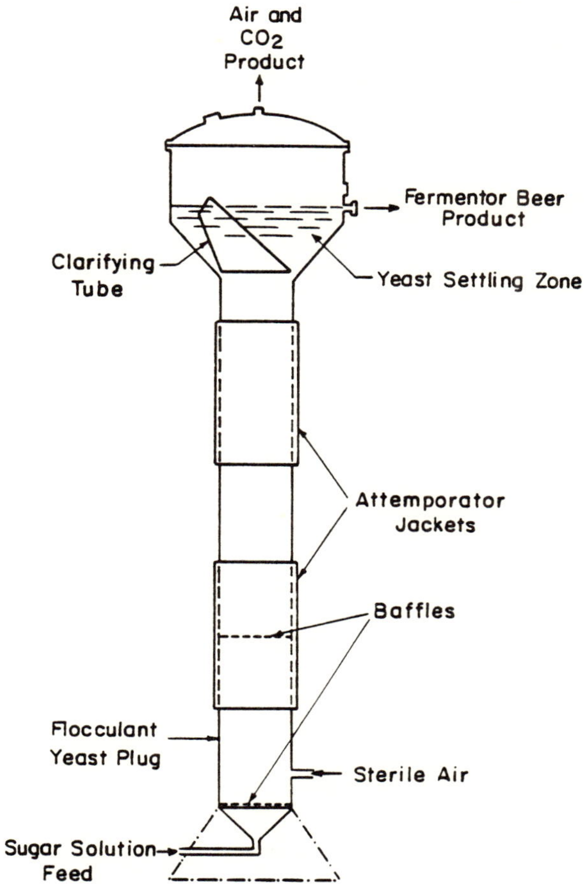

APV Tower Fermentor

Figure V-9.

A major drawback of the APV system is the long time required for initial start-up. Two to three weeks are required to build up the desired high density and achieve stable operation (2). This is compensated by the very long, 12 months and greater, on times between shutdowns.

Based on the simple process selection criteria the tower fermentor again appears to be an improvement over the processes discussed earlier. The system is very simple. While added power is required for pumping the beer up through the tower, this is offset by the savings from the elimination of a fermentor agitator. The high productivity achieved in the APV system results in very small fermentors. Capital and operating costs are predicted to be far lower than for conventional batch processes.

The slant tube fermentor has been proposed as a modification to the tower fermentor with a higher beer flow rate capacity (53). Three cm diameter tube 14 meters long is mounted at a 45 degree angle to horizontal. Sugar solution is pumped through the tube from the base. Fermentation takes place progressively up the tube. Retention of yeast in the slant tube at high flow rates is possible as the yeast settling depth is very shallow--yeast need only settle to the tube lower wall. Once settled, yeast roll rapidly down the tube. As CO_2 is evolved it rises to the tube upper wall where it bubbles rapidly up and out of the fermentor. A three phase flow results with CO_2 flow separated and therefore no longer interfering with yeast settling. In a laboratory study a fermentor productivity of 25 g/L-hr was achieved. This rate is 39 times greater than that for a simple batch fermentation with the same grape juice feed (53).

A bank of many slant fermentor tubes operated in parallel would be required to achieve industrial scale production. Laboratory tests have shown that a single pump feeding several parallel tubes from a single manifold could not maintain equal flow rates in them all (53), and a special flow distributor system will be required for an industrial fermentor.

While the slant tube fermentor may allow somewhat increased productivity over the simple APV tower design, the added complexity of a multiple tube system will probably offset this advantage.

Another approach to allowing increased throughput in tower fermentors has been to pack the fermentor with standard distillation column type packing (54). The packing material provides zones of quiescence where yeast cells can collect and very high cell densities can be maintained even at high flow rates. Continual cell growth can cause selective plugging of the fermentor and fluid by-passing so that the column must be frequently sparged with a rapid nitrogen flow to shake free and redistribute the cells (54). This disrupts stable continuous operation and may mitigate the advantages of slightly increased productivity.

5. Dialysis Fermentors

Dialysis fermentation was first developed as a batch technique, using a simple dialysis flask fermentor (55,56). This process is readily adapted to continuous culture (Fig. V-10)(30).

In the simple continuous dialysis system shown, a culture of actively fermenting yeast is maintained in a confined zone of the fermentor. Substrate enters the fermentation zone by diffusing through a dialysis membrane from the medium zone. Fermentation then takes place and product ethanol diffuses back through the membrane into the medium zone where it is recovered in an overflow. Yeast cannot escape the fermentation zone so that extremely high cell densities can be achieved.

For this system, fermentation rate is limited by the rate at which substrate can diffuse across the limited membrane surface area, not by the

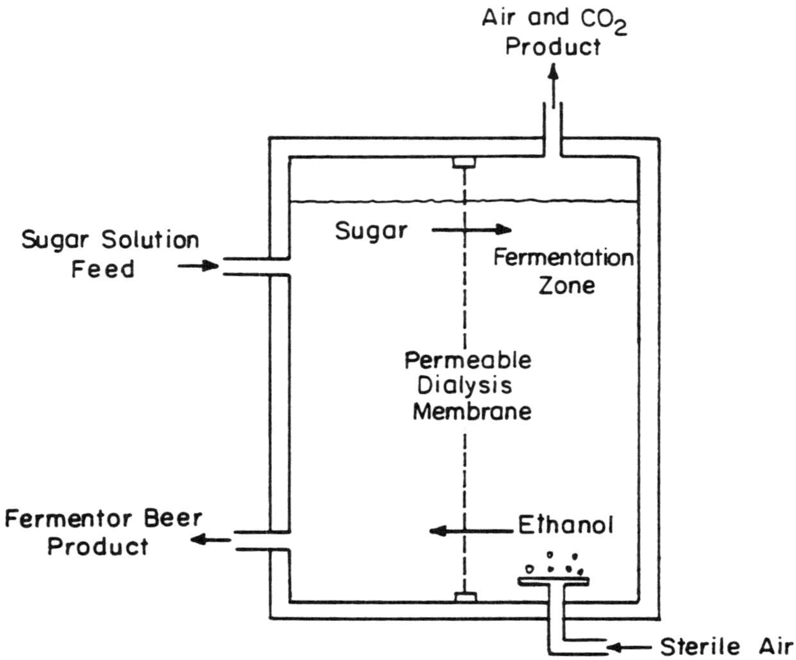

Simple Continuous Dialysis Fermentor

Figure V-10. Production Rate Limited by Substrate Diffusion to the Fermentation Zone.

inherent metabolic limitations discussed earlier. This practical diffusional limitation prevents very rapid fermentation in a simple system like that shown and thus makes it impractical for industrial application to ethanol production.

The limitation of slow substrate diffusion rate can be overcome if pressure dialysis is used (Fig. V-11). Here, a pressure differential is applied across the dialysis membrane forcing a bulk flow of medium through the well-stirred culture zone. This system is also impractical because free proteins (the products of cell lysis) rapidly pile up on the membrane building up a thick cake which blocks the membrane pores and prevents further flow (57).

The problem of membrane fouling has been overcome in the rotorfermentor (40) (Fig. V-12)(58).The rotorfermentor is a continuous pressure dialysis reactor. The simple fixed membrane is replaced by a rapidly rotating membrane cylinder. Feed and air are pumped into the annular fermentation zone where yeast grow and are retained. This zone is maintained under pressure (2-20 psig)(59), so that filtrate continuously flows out through the membrane. Because the membrane is rotating, a strong centripetal force is developed at the membrane surface and large molecules impinging on the surface are thrown back into the annular zone. Only a very thin steady state cake thickness is developed. High filtration rates, 3.4 L/hr for a 1020 cm^2 membrane surface, and 2 psig pressure drop can be maintained. Using a laboratory scale rotorfermentor cell densities of 30.9 g/L and ethanol productivities of 36.5 g/L-hr have been achieved (59).

The rotorfermentor appears attractive in terms of its high productivity. It fails, however, when rated on mechanical and operating simplicity. The rotating membrane unit is mechanically complicated and seals for this unit

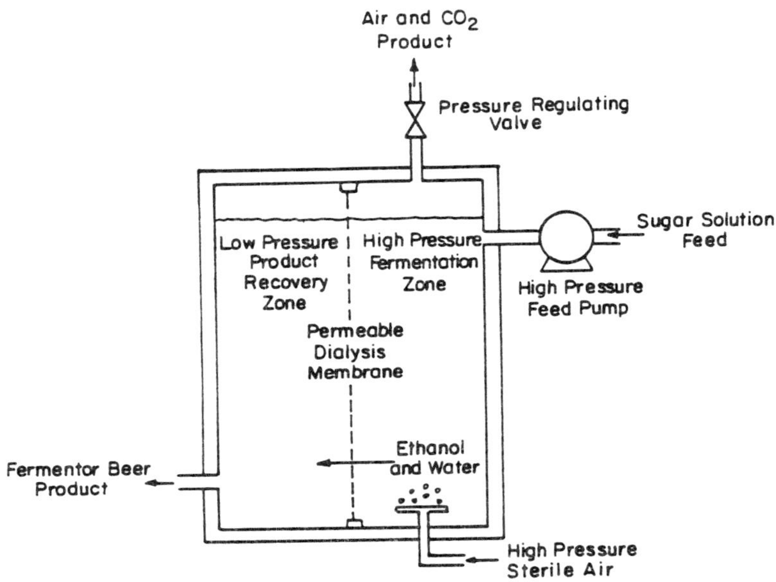

Pressure Dialysis

Figure V-11.

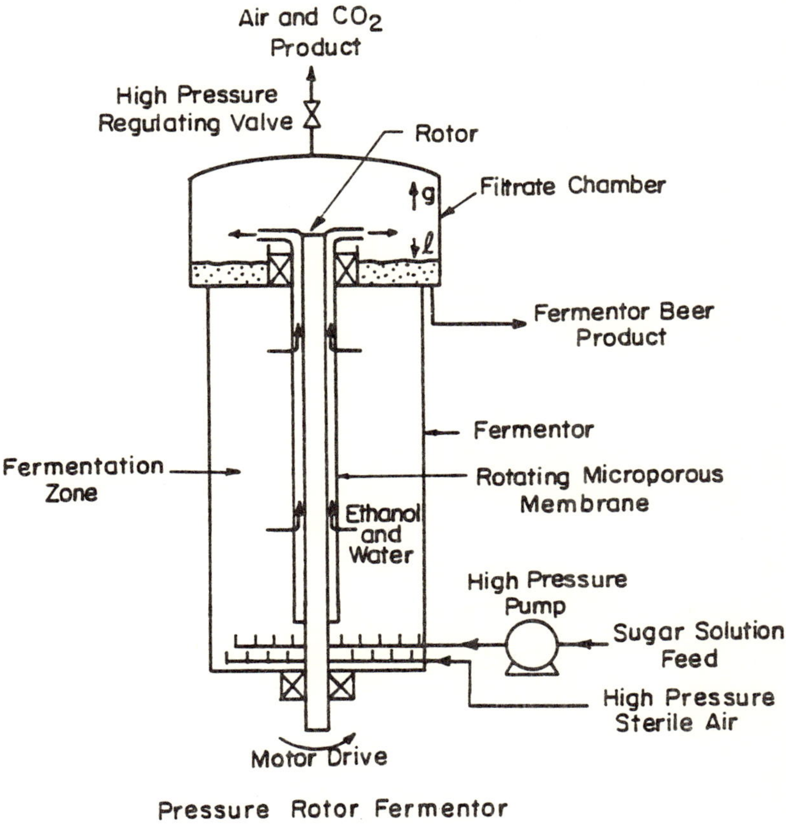

Figure V-12. Hollow Fiber Fermentor. Feed Flow and Product Recovery is through the Fiber Membrane Bundle. Fermentation Occurs in the Shell Side High Yeast Density Zone.

will be made especially complex as they must not leak against the fermentor pressure. More important though will be the requirement to periodically replace the membrane. The long term mechanical stability of molecular filtration membranes under high shear has not been studied, but the developers of the rotorfermentor agree that periodic replacement will be necessary (60). Membrane replacement will interrupt continuous operation and will require considerable skilled labor time. Because of these complications the rotorfermentor probably will not become a competitive industrial process despite its high productivity.

Another approach to achieving high rate dialysis fermentation involves the use of hollow fiber reactors (Fig. V-13)(61).Hollow fiber reactors provide extremely large membrane surface areas so that rapid substrate diffusion and high fermentation rates are possible. The hollow fiber reactor is arranged like a shell in tube heat exchanger. The tubes are fine hollow fibers (0.020 inch I.D.) of membrane material (62). A bundle of 1000 of these fiber tubes can be packed into a single three cm diameter shell (63). The shell side is then inoculated with growing cells. Medium is fed into the fiber tubes at one end of the reactor and glucose substrate diffuses out and is fermented. Ethanol product diffuses back into the tubes and is carried out at the far end of the reactor.

Hollow fiber reactors have been used at laboratory scale to produce several high value biological products (61,64,65). Work utilizing hollow fiber reactors for ethanol production is just beginning at Berkeley (66) and Stanford (67).

The hollow fiber ractor has potential to achieve yeast cell densities approaching the maximum cell packing density (155 g/L) (68) and corresponding high ethanol productivities should be possible. Hollow fiber reactors are

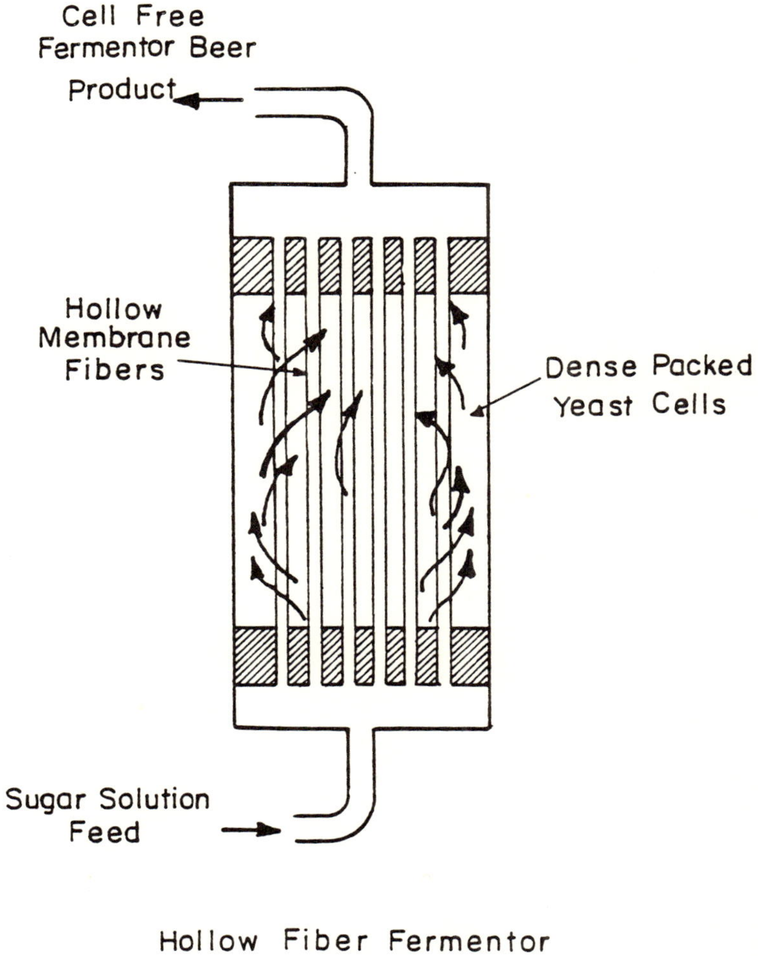

Figure V-13. Hollow Fiber Fermentor. Feed flow and product recovery is through the fiber membrane bundle. Fermentation occurs in the shell side high yeast density zone.

quite complex and very costly (63). Membrane plugging may also be a problem. If these problems are not overcome, the hollow fiber reactor probably will not be important as an industrial ethanol fermentation reactor.

The plug fermentor (Fig. V-14) is a particularly interesting version of the dialysis fermentor (69). A dense plug of yeast is maintained between two support plates. Medium is pumped through the plug under high pressure (0.2 to 10 atmospheres) and rapid fermentation takes place (70).

Easily fouled membrane filters are not used. Instead a layer of Kieselguhr filter aid over a porous frit support plate retains the yeast plug (71). Kieselguhr is also mixed into the yeast plug to prevent dense packing of the yeast which would halt the flow.

Using this system, productivities 72 times greater than for simple batch fermentation with the same yeast and substrate have been achieved (69).

Oxygen concentration varies considerably across the plug. Yeast near the feed receive air saturated medium, but yeast near the product end receive oxygen depleted medium. Cell viability decreases with time because of this and other limitations and after 27 days operation, cell viability is only 52%. Productivity experiences a similar decline to 50% of the maximum level. Regular shutdown and rejuvenation of the yeast under ordinary growth conditions will probably be required for sustained industrial operation.

The high productivity of the plug fermentor makes it quite attractive. The requirement of high pressure equipment will partly offset the capital cost savings resulting from reduced vessel volume. High power pumps will be required to force medium through the plug and the required shutdowns

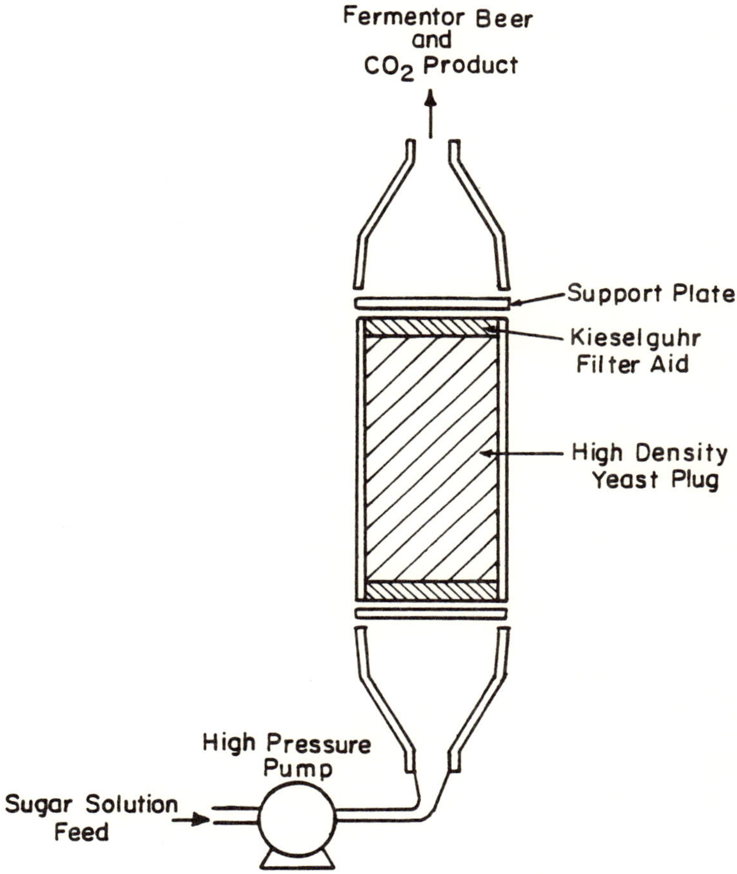
Fermentor Beer
and
CO_2 Product
Support Plate
Kieselguhr
Filter Aid
High Density
Yeast Plug
High Pressure
Pump
Sugar Solution
Feed
Plug Fermentor

Figure V-14.

for yeast regeneration will interrupt continuous operation. Further study is required to determine if these disadvantages are offset by the very high productivities possible.

6. Bound Cell Fermentors

Yeast cells can be immobilized by entrapment in a gel matrix or by covalent binding to surfaces (72).

Gel entrapment has been studied by many researchers (73,74). In one study, yeast cells were suspended in 1% sodium alginate solution. This mix was then slowly extruded into .05 M $CaCl_2$ to produce fine fibers in which the yeast were trapped. These fibers were then packed into a column reactor: 10% glucose solution was forced through the bed and fermentation proceeded giving a 90% yield in ten hours residence time (73).

This fermentor system is especially simple as no agitation or yeast recovery equipment is required. It is found, however, that yeast viability declines for alginate gel entrapped cells with a cell half life at only ten days (73). Frequent gel fiber replacement would therefore be required to maintain high productivity.

Recent work using polyacrylamide gels suggest that the yeast viability problem can be overcome (75). If high productivities can be maintained in an industrial gel-entrapment fermentor, this could be a very attractive industrial process.

Another cell retention technique involves binding cells to solid surfaces. A thick gel can impose mass transfer limitations which slow the fermentation rate (76). These limitations are reduced if very thin gel layers on surfaces are used and are virtually eliminated if the cells are directly chemcially bound to surfaces.

In one immobilization technique, ordinary distillation column packing is coated with a suspension of yeast cells in a polyelectrolyte. The polyelectrolyte is then crosslinked into a thin film gel with glutaraldehyde (77). A laboratory column fermentor packed with this material achieved an ethanol productivity of 35 g/L-hr.

Yeast cells have been bound to cotton gauze using cyanuric chloride to form a covalent bond between the exposed cotton hydroxyl groups and cell wall proteases or polysaccharides (76). Ligand bonding to metal hydroxides has also been used (78). Both of these techniques did result in cell retention and in both cases viability was maintained for long periods after immobilization. Further study is necessary to determine maximum productivities obtainable with chemically bound cells before any evaluating of the potential of this technique can be made.

7. Extractive Fermentation

The fermentation techniques describe thus far have achieved high productivities by maintaining high cell densities. Now to be considered are techniques which can also achieve high specific ethanol productivities by maintaining low ethanol concentrations in the fermentor and thus eliminating end product inhibition.

In the extractive fermentation process (Fig. V-15) ethanol is continuously removed from the fermentor beer by solvent extraction (79). A side stream of beer is tapped from the fermentor, the cells are removed and recycled, and the clear beer is contacted with a liquid extractant. The extractant, which is immiscible in the beer, absorbs most of the ethanol. The purified beer can now be returned to the fermentor and the ethanol saturated extractant can be processed through a distillation train for ethanol recovery and solvent recycle.

To be successful, an extractant must have these properties (80):

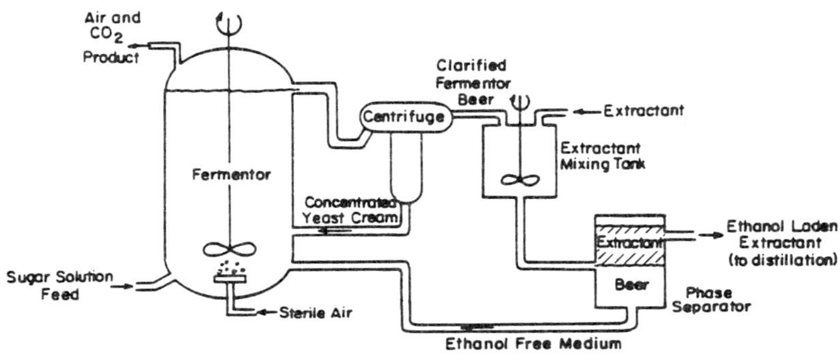

Continuous Solvent Extraction Fermentation

Figure V-15.

1. non-toxic to yeast
2. high distribution coefficient for ethanol
3. selective for ethanol over water and secondary fermentation products.
4. should not form emulsions with fermentation broth.

Extractive fermentation has the potential to offer very high fermentation rates by eliminating ethanol inhibition. If in addition to the required properties the extractant is of low volatility, distillation will be simplified and considerable energy savings could result over conventional systems.

In terms of the process evaluation criteria, extractive fermentation with little increase in complexity yielding a simple very high productivity low energy fermentation system. Unfortunately, no extractant having all the required properties has thus far been found (81).

8. Product Recovery Membrane Fermentation

Continuous ethanol recovery from the fermentor broth can also be achieved using selective membrane separation techniques.

Membrane extractive fermentation (Fig. V-16)(81) is similar to simple extractive fermentation except that the extractant is separated from the fermentor broth by a diffusion membrane. Ethanol being more soluble in the extractant, diffuses across the membrane and is carried away. Membrane material which will allow ethanol removal while retaining sugar substrate in the fermentor can be used, and this helps assure complete substrate utilization (82).

Requirements for the extractant are far less severe for use in membrane extractive fermentation than in direct extractive fermentation. Only a very small amount of extractant--that which leaks through the membrane--

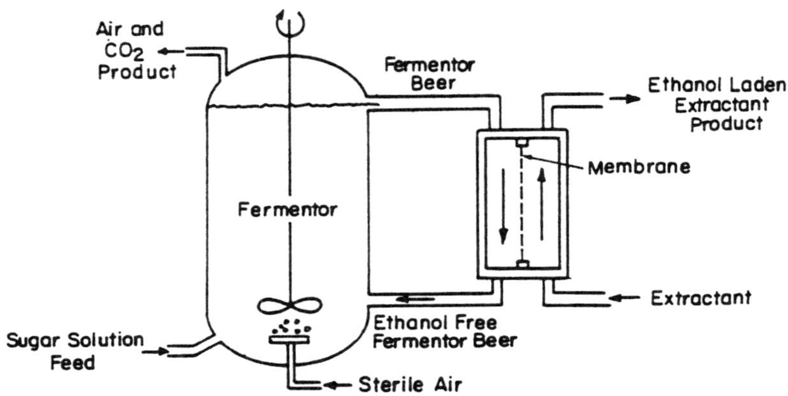

Continuous Membrane Extractive Fermentation

Figure V-16.

contacts the broth, and the requirements of nontoxicity can thus be reduced. The requirements of immiscibility and non-emulsion forming properties can also be reduced. The extractant polypropylene glycol p-1200 has been identified as suitable, having a distribution coefficient of 0.60 and being only slightly toxic to yeast (83).

Like simple extractive fermentation, membrane extractive fermentation appears promising. The process is simple, involves little added equipment and may again allow an energy reduction in distillation. Fouling of the membrane is a possible drawback as this would require frequent costly shut-downs for membrane replacement. Studies to determine maximum fermentation rates achievable with membrane extractive fermentation are now underway (80).

Selective product recovery membrane fermentation is another new technique under development. The apparatus used is like that used for membrane extractive fermentation, except that no extractant is used, The membrane itself performs the separation, facilitating ethanol diffusion through the membrane while retarding water and other beer components. Selective ultra-filtration membranes capable of maintaining the fermentor ethanol concentration at less than 60 g/L while yielding a product concentration of 120 g/L have already been developed and membranes capable of maintaining beer concentrations at below 20 g/L should be available soon (84). Flux rates across these membranes are still quite low and large membrane surfaces are required (85). Membrane fouling may again be a problem. If these difficulties can be overcome so that continuous operation could be assured without frequent membrane replacement, so that membrane cost would not be prohibitive, then this process would be attractive for industrial use.

9. Vacuum Fermentation

The Vacu-ferm process (Fig. V-17) developed concurrently by Cysewski and

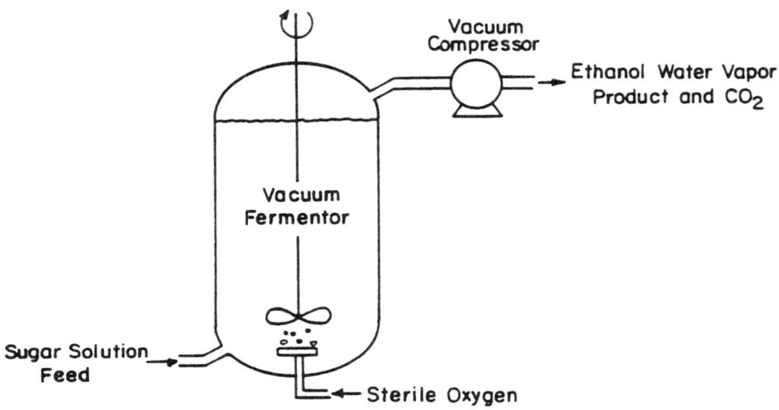

Continuous Vacuum Fermentation

Figure V-17.

Wilke (28) and Ramalingham and Finn (86) provides an alternative for fermentation with continuous ethanol removal. A concentrated sugar solution is fed continuously to the reactor. Fermentation is conducted under vacuum and an ethanol water solution is boiled away, maintaining the liquid level constant. Since ethanol is more volatile than water, then this flashing operation acts as a single stage distillation. With a fermentor temperature of 35°C and pressure of 51 mmHg, the ethanol concentration in the fermentor is maintained at 35 g/L while the ethanol concentration in the vapor product is increased to more than 200 g/L, thus simplifying later distillation steps (7). Yeast cells build up in the fermentor and cell densities of 120 g/L can be maintained. With ethanol inhibition removed a total productivity of 80 g/L-hr can be achieved (3).

There has been some concern over the added energy requirements to drive vacuum compressors for the vacu-ferm process (43). It has been shown, however that energy requirements for the vacu-ferm process are increased only five percent over these for conventional processes when suitable techniques for energy recovery are employed (87).

The absolute productivity achieved in the vacuum fermentor is higher than has yet been achieved in any other device and this advantage far outweighs the small increase in energy requirements for vacuum compressor operation. Other difficulties are associated with vacu-ferm operation, however. To meet the yeast oxygen maintenance under vacuum, pure oxygen must be sparged through the fermentor, and this contributes an added cost of 2¢ per gallon to the cost of alcohol produced (7). The compressors required must operate at unusually low pressures and are extremely large. They will be costly and difficult to control (88). Vacuum operation will increase the likelihood of fermentor contamination and shutdown. It is not clear whether the outstanding productivity of the vacuum fermentor can conpensate for these potential operating difficulties. Pilot plant operation has been proposed (89), however, and should resolve this question.

The flash-ferm process modifies the simple vacu-ferm process to overcome some of its operating difficulties (Fig. V-18)(7). Fermentation is carried out in an atmospheric pressure fermentor so that the yeast oxygen maintenance requirement can be cheaply met with sparged air. Carbon dioxide produced in the fermentation process can be directly vented from the fermentor and thus needs not be compressed with the vapor product as in vacu-ferm. To remove ethanol, beer is rapidly cycled through a small auxiliary flash vessel where it boils. Ethanol is recovered as the flash vessel overhead vapor product and ethanol depleted beer is returned to the fermentor for further fermentation. The problem of contamination is greatly reduced as only the small flash vessel is under vacuum. Energy requirements for this process are also slightly lower than for vacu-ferm.

The flash-ferm plant is somewhat more complicated than the vacu-ferm plant, the former requiring and added vessel and the associated liquid cycling pumps, but this is probably more than compensated by the advantages cited. Like the direct vacu-ferm process, it is not clear whether the high productivities possible with vacuum operation offset the operation and control difficulties.

C. Conclusion

Twenty different advanced high rate fermentation processes have been described and assessed as to their potential for application in industrial alcohol production. The comparison made are summarized in Table V-2. Clearly many alternatives superior to the conventional batch technology of the 1930's

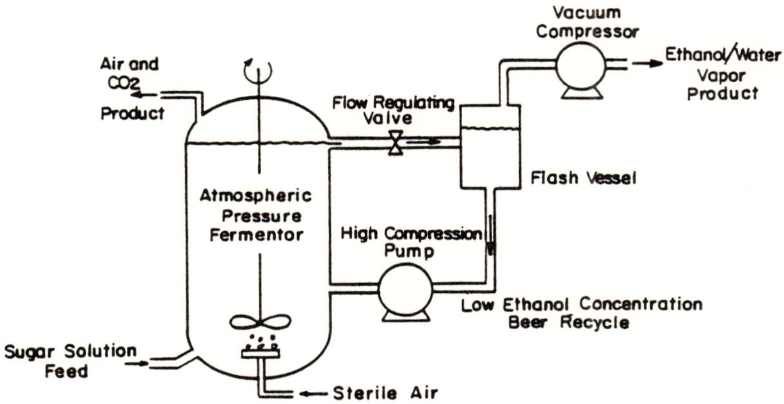

Continuous Flash Fermentation

Figure V-18.

Table V- 2

Alternative High Rate Fermentation Processes

Process	Fermentation Productivity	Comments	References
Batch Fermentation	Very Low (1.8 to 2.5 g/L-hr.)	Very high capital and operating cost.	4,5,6,7,8.9
Simple CSTR Fermentor	Low (∿ 6 g/L-hr)	Mechanically simple equipment, simple continuous operation.	6,28,37
Series CSTR Fermentors	2-3 times rate for simple CSTR	Simple continuous operation.	2,25,34,35,39
Perforated Plate Column Fermentor	Not reported	Mechanically complex fermentation device	90,91,92,93,94
CSTR with Centrifuge Cell Recycle	High (30-40 g/L-hr.)	Added energy requirements and added operator attention required for centrifuge.	3,6,9,24,25 40
CSTR with Alternative Recycle Scheme	High	Settlers appear to be too large for industrial application. Whirlpool separators may be attractive.	4,46,47
Tower Fermentors	High	Mechanically simple fermentor with simple continuous operation, but start up period is very long.	2,48,49,50,51,52
Slant Tube Fermentors	High	Mechanically complex	53
Packed Bed Tower Fermentor	High	Plugging and by-passing are major problems.	54
Dialysis Fermentor	Not reported	Fermentation rate limited by substrate diffusion through membrane. Membrane fouling may require frequent shutdowns.	30,55,56

Table V-2 Continued

Process	Fermentation Productivity	Comments	References
Pressure Dialysis Fermentor	Not reported but potentially high	Membrane fouling will be a major problem.	57
Rotor Fermentor	High (36 g/L-hr)	Membrane fouling problem overcome but the fermentor device is mechanically quite complex. Membrane distruction by mechanical shear may be a problem.	24,58,59,60
Hollow Fiber Fermentor	Not reported but potentially high	Expensive fermentation units. Mass transfer limitation may limit productivity.	61,62,63,64, 65,66,67,68
Plug Fermentor	High (72 times greater than for similar batch fermentation)	Frequent shutdowns for yeast rejuvenation in an aerobic environment will be required. High power feed pumps are required.	69,70,71
Gel Entrapment	90% yield from 10% glucose solution in ten hours	Very simple fermentor system with no agitation or recycle equipment required. Half life of entrapped cells must be increased to make this process attractive.	73,74,75
Chemically Bound yeast cells	Not reported	The cell viabilities do remain high after covalent bonding. If productivties also remain high, this could yield a very simple, high productivity fermentation system.	72,76,77, 78
Direct Extractive Fermentation	Potentially very high	No suitable extractant has been found.	79,80,81
Membrane Extractive Fermentation	Not reported, but potentially very high	Membrane fouling may be a problem. Process is otherwise simple.	80,83

Table V-2 Continued

Process	Fermentation Productivity	Comments	References
Selective Membrane Fermentation	Not reported but potentially very high.	Membranes with suffficiently high throughput rates have not yet been developed.	84,85
Vacuum Fermentation	Very high (80 g/L-hr.)	Mechanically complicated equipment requiring constant monitoring. Small added energy requirements for vapor compression. Contamination of the vacuum fermentor may be a problem. Pure oxygen must be sparged to the fermentor.	3,7,28,43 86,87,88,89
Flash Fermentation	Very high	Mechanically very complicated. Energy requirements only slightly increased over conventional processes. Contamination problem greatly reduced as compared to direct vacuum fermentation and pure oxygen is not required.	7

exist. Among these, simple continuous, series continuous, cell recycle and APV tower fermentors have been operated at large scale with considerable savings over batch processes. These should then be the processes of choice for any new ethanol plants.

Among processes with continuous ethanol removal to eliminate end product inhibition, only vacuum and flash fermentation have been advanced sufficiently to allow pilot plant testing. The potential advantages of these processes in super high productivities certainly warrants this next stage in their evaluation.

The other high rate processes, dialysis fermentation, bound cell fermentations and extractive and selective membrane fermentations require much more evaluation before their merits can be fully assessed.

References

1. Chem. Eng. News 51, #10, 7 (1973).
2. J.S. Hough, C.W. Keevil, V. Marie, G.P. Kirk, and T.W. Young, "Continuous Culture in Brewing," Chapter 17.
3. G.R. Cysewski, "Fermentation Kinetics and Process Economics for the Production of Ethanol," Ph.D. Thesis, Univ. of California, Dept. of Chemical Engineering, Berkeley, LBL-4480, March (1976).
4. P.B. Jacobs "Industrial Alcohol," U.S.D.A., Washington, D.C., Misc. Publication #695, Feb. 1950,
5. D. Rose, "Yeast for Molasses Alcohol," Process Biochemistry 11, #3, 10 (1976).
6. G.D. Cysewski, and C.R. Wilke, Biotech. and Bioeng. 20, 1421 (1978).
7. B. Maiorella, H.W. Blanch and C.R. Wilke, "Rapid Ethanol Production via Fermentation," Univ. of California., LBL-10219, Nov. (1979). Presented at AIChE 72nd meeting, San Francisco, CA., November 29, 1979.
8. R. Katzen Associates, "Grain Motor Fuel Alcohol Technical and Economic Assessment Study," Cincinnati, Ohio, U.S.D.O.E. contract #EJ-78-C-01-6639, June (1979).
9. W. Fong, J. Jones and K. Semrau, "The Cost of Producing Fermentation Ethanol from Biomass," SERI International, presented at AIChE 72nd Meeting, San Francisco, CA, November 29, 1979.
10. Reference 3, page 9
11. S.J. Pirt, W. Kurowski, J. Gen. Microbiol. 63, 357 (1970).
12. H. Holzer, "Aspects of Yeast Metabolism," A.K. Mills Ed., Oxford Blackwells Scientific Publications (1968).
13. M.D. Akbar, P.D. Richard, and F.J. Moss, Biotech. and Bioeng. 16, 455 (1974).
14. F.J. Moss, P.D. Richard, F.E. Bush, Biotech. and Bioeng. 13, 63 (1971).

15. P.D. Richard, F.J. Moss, D. Phillips, T.C.K. Mok, Biotech & Bioeng. 13, 164 (1971).

16. P.J. Rogers, P.R. Stewart, J. Gen. Microbiol. 79, 205 (1973).

17. A.D. Haukeli, S. Lie, J. Inst. Brew. 79, 55 (1973).

18. M.H. David, B.H. Kirsop, J. Inst. Brew 79, 20 (1973).

19. D.K. Bloomfield, K.J. Black, Bio Chem. 235, 337 (1960).

20. J. White, D.J. Munns, Vallestein Commun., 14, 199 (1951).

21. T.W. Cowland, D.R. Maule, J. Inst. Brew. 72, 480 (1966).

22. A.A. Andreasen, J.J.B. Stier, J. Cell. Comp. Phys. 41, 23 (1953).

23. A.S. Aigar, R. Ludeking, Chem. Eng. Prog. Symp. 62 (69), 57 (1969).

24. C. Bazua, "Effect of Alcohol Concentration on Kinetics of Ethanol Production by Saccharomyces cerevisiae," M.S. Thesis, Univ. of Calif., Dept. of Chem. Eng., Berkeley (1975).

25. T.K. Ghose, R.D. Tyagi, Biotech. & Bioeng. 21, 1401 (1979).

26. S. Aiba, M. Shoda, M. Nagatani, Biotech. & Bioeng. 10, 845 (1968).

27. I. Holzberg, R. Finn, K. Steinkraus, Biotech. & Bioeng. 9, 413 (1967).

28. G. R. Cysewski, C.R. Wilke, Biotech. & Bioeng. 19, 1125 (1977).

29. E. Delrosario, K. Joon Lee, P. Rogers, Biotech. & Bioeng. 21, 1477 (1979).

30. S. J. Pirt, "Principles of Microbe and Cell Cultivation," Halsted Press (1975).

31. I.A. Webster, M.L. Shuler, Biotech. & Bioeng. 21, 1725 (1979).

32. T.W. Nagodawithana, C. Castellano, K.H. Steinkraus, Appl. Microbiol 28 383 (1974).

33. T.W. Nagodawithana, K.H. Steinkraus, Appl. Environ. Microbiol. 31, 158 (1976).

34. J. S. Hough, "Production of Beer by Continuous Fermentation," Brewing Industry Research Foundation, Nutfield, Surrey, England (1959).

35. V.L. Yarovenko, "Theory and Practices of Continuous Cultivation of Microorganisms in Industrial Alcoholic Processes," Advances in Biochem. Eng. Vol 9, T.K. Ghose, A. Fiechter, N. Blakebrough, Editors, Springer-Verlag, Berlin (1978).

36. P.S.S. Dawson, "Continuous Fermentations," Chapt. 4, Annual Reports on Fermentation Processes, Vol. 1, D. Perlman, Editor, Academic Press, N.Y. (1977).

37. O. Levenspiel, "Chemical Reaction Engineering," John Wiley & Sons, New York (1962), Chapter 5.

38. Ref. (3) p. 133

39. L. Bishop, J. Inst. Brew. 76, 172 (1970).

40. L.D. Sortland, "The Kinetics of Dense Culture Fermentations," Ph.D. Thesis, Univ. of California, Dept. of Chem. Eng., Berkeley, May (1968).

41. S. Elias, "Gasohol, New Role for Food Industry," Food Engineering, Oct. (1979), p. 61-65.

42. A.D. Portno, J. Inst. Brew. 73, 43 (1967).

43. T. Ghose, R.D. Tyagi, Biotech. & Bioeng., 21, 1387 (1979).

44. T. Wash, H. Bungay, Biotech. & Bioeng. 21, 1081 (1979).

45. A. Kitai, R. Okamoto, A. Ozaki, "Continuous Culture using a Perforated Plate Column," Proc. IV Int. Ferm. Symp., Kyoto, Japan, March 19 (1972).

46. F. Jirmann, U.D. Runkel, Brauwelt, 107, 1453 (1967).

47. J. Hough, P. Gough, A. Davis, J. Inst. Brew. 68, 478 (1962).

48. R. Greenshields, E. Smith, "Tower Fermentation Systems and Their Applications," Chem. Engineer, #249 May (1971).

49. M.G. Royston, Process Biochem. 1, 215 (1966).

50. W.J. Klopper, R.H. Roberts, M.G. Royston, R.G. Ault, Proc. of X Cong. of Eur. Brew. Convention, Stockholm, 1965, p. 238. (Elsevier Pub. Co., Holland).

51. D.T. Shore, M.G. Royston, Chem. Eng. Lond. 218, CE 99 (May 1968).

52. R. G. Ault, A.N. Hampton, R. Newton, R.H. Roberts, J. Inst. Brew. 75, 260 (1969).

53. E. Wick, K. Popper, Biotech. & Bioeng. 19, 235 (1977).

54. O.C. Sitton, J.L. Gaddy, "Design and Performance of an Immobilized Cell Reactor for Ethanol Production," AIChE, 72 Ann. Meeting, San Francisco, November 25 (1979), #41D.

55. P. Gerhardt, D. Gallup, J. Bact. 86, 919 (1963).

56. J. Schultz, P. Gerhardt, Bact. Rev. 33, #1 (1969).

57. B. Maiorella, T. Glenchur, Univ. of California, Dept. of Chem. Eng., Berkeley, Notebook, RB 300 (1979).

58. A. Margaritas, C.R. Wilke, Biotech. & Bioeng. 20, 709 (1978).

59. A. Margaritas, C.R. Wilke, Biotech. & Bioeng. 20, 727 (1978).

60. Communication with Prof. C.R. Wilke, Univ. of Calif. , Dept. of Chem. Eng., Berkeley (1979).

61. J.K. Kan, M.L. Shuler, "An Immobilized Whole Cell Hollow Fiber Reactor or Urocanic Acid Production," AIChE Symp. Ser. #172, Food Pharm. And Bioeng. (1976-1977) p. 31.

62. B. R. Breslau, B.M. Kilcullen, "Continuous Ultrafiltration," Amer. Dairy Sci. Assoc., 71 Ann. Meeting, Raleigh, N.C., June 21 (1976).

63. "Romicon Hollow Fiber Ultra Filtration Membrane Cartridges," Romicon, Inc., 100 Cummings Park, Woburn, Mass. 01801 (1979).

64. I.A. Webster, M.L. Shuler, P.R. Rony, Biotech. & Bioeng. 21, 1725 (1979).

65. J.K. Kan, M.L. Shulter, Biotech. & Bioeng. 20, 217 (1978).

66. Communication -- Bruce Vick Roy, Univ. of Calif., Dept. of Chem. Eng., Berkeley (1979).

67. C. R. Robertson, "Enzymes, Microbes, and Chemical Engineers," Chem. Eng. Dept., Stanford Univ., presented at Univ. of Calif., Berkeley, Chemical Engineering Colloquia, Jan. 21 (1980).

68. Communication -- Prof. H.W. Blanch, Univ. of Calif., Dept. of Chem. Eng., Berkeley (1979).

69. D. Baker, B.H. Kirsop, J. Inst. Brew. 79, 487 (1973).

70. P. Berdelle - Hilge, U.S. Patent # 3,737,323, June 5 (1973).

71. "Chemicals by Yeast Fermentation," p. 259 - 289, Food Tech. Rev., #45 (Yeast for Food and Other Purposes) Jeanne C. Johnson, Ed., Noyes Data Corp., Parkridge, N.J. (1977).

72. B. Abbot, "Immobilized Cells," Chapt. 5, Annual Reports on Fermentation Processes, Vol. 2, D. Perlman Ed., Academic Press, New York (1977).

73. M. Kierstan, C. Bucke, Biotech. & Bioeng. 19, 387 (1977).

74. C. Divies, French Patent #844,766 (1977).

75. I. Chibata, T. Tosa, T. Sato, Appl. Microbiol. 27, 878 (1974).

76. K. Venkatasubramian, et al., "Enzyme Engineering," #3, L.K. Pye and H. Weetal, Eds., Plenum Press (1977).

77. W. Griffith, A. Compere, Dev. Ind. Microbiol. 17, 241 (1976).

78. J. Kennedy, S. Barker, J. Humphreys, Nature 261, 242 (May 20, 1976).

79. E.K. Pye, A.E. Humphrey, "The Biological Production of Liquid Fuels from Biomass," Univ. of Penn., Interim report to U.S.D.O.E., Task 7, (June - August, 1979) p. 79-86.

80. E.K. Pye, A.E. Humphrey, "Production of Liquid Fuels from Cellulosic Biomass," Proc. - 3rd Ann. Biomass Energy Systems Conf., p. 69, U.S.D.O.E., Solar Energy Res. Inst., Golden, Colo., (June 5, 1979).

81. E.K. Pye, A.E. Humphrey, "The Biological Production of Liquid Fuels from Biomass," Univ. of Penn., Interim Report to U.S. D.O.E., (Feb. - May 1979).

82. L. Craig, "Differential Dialysis," Science 144, 1093 (May 29, 1964).

83. Ref. 70- Task 8.

84. H. Gregor, "Membrane Processes of Separation and Concentration in Biomass Harvesting, Production and Refining," Proc. 3rd Ann. Biomass Energy System Conf., p.-39, U.S. D.O.E. Solar Energy Res. Inst., Golden, Colo. June 5, 1979).

85. H. Gregor, T. Jefferies, "Ethanolic Fuels from Renewable Resources in The Solar Age," Annls. of New York Academy of Science, Vol. 385.

86. A. Ramlingham, R.K. Finn, Biotech. & Bioeng. 19, 583 (1977).

87. B. Maiorella, C.R. Wilke, "Energy Requirements for the Vacu-Ferm Process," Biotech. & Bioeng. 22, 1749-1751 (1980).

88. A. Zimmer, "Latest Developments in Energy Efficient Evaporation," Weigand Evaporators, Inc., 5585 Sterrett Place, Columbia, Maryland.

89. D.J. O'Neil, A.R. Colcord, M.K. Bery, R.S. Roberts, D. Sondhi, S.Day, I.A. Barbary, "Design, Fabrication and Operation of a Biomass Fermentation Facility," Tech. Prog. report #2 to U.S. D.O.E., Georgia Inst. Tech., Atlanta (Jan-March 1979) p. 112.

90. E. Falch, E. Gaden, "A Continuous, Multistage Tower Fermentor. I. Design and Performance Tests," Biotech. & Bioeng., 11, 927 (1969).

91. A. Ktai, R. Okamoto, A. Ozaki, "Continuous Culture Using a Perforated Plate Column," Fermentation Technology Today, G. Terni, Ed. (1972).

92. A. Kitai, T. Yamagata, "Perforated Plate Column Fermentor," Process Biochem., Nov. 1970.

93. J. Paca, V. Gregr, Biotech. & Bioeng. 19, 539 (1977).

94. A. Prokop, L. Erickson, J. Fernamdez, A. Humphrey, Biotech. & Bioeng. 945 (1969.)

VI. Ethanol Economics

A. Cost Summaries for Major Processes

Only Natick, Berkeley, Gulf/Arkansas, M.I.T. and GE/CRD of all the processes considered have defined processes sufficiently well to publish cost analyses. The cost analyses generated independently by each of these groups are summarized in Table VI-1. It must be stressed that the bases of these analyses are not entirely consistent, and more details on the assumptions for these estimates would be required to allow a direct comparison.

Furthermore, none of the processes have been demonstrated on a large enough scale to justify all the assumptions made in the cost estimates for commercial production of ethanol from cellulose. Because both Natick and Gulf/Arkansas have operated their processes at least on a pilot plant scale, they should have more reliable bases for their assumptions. However, it is not clear whether these processes have actually been run at the conditions and length of time called for in the design of their commercial production processes. Other questions with regard to their cost estimates are the availability of cellulose feedstocks at the assumed prices and the apparently high quantity of yeast solids being claimed as feed credits (1).

The Berkeley process design is based on bench scale data. The cost estimate for this process is discussed in detail in the following section. The basis for the GE/CRD cost estimate and the Battelle cost estimate of the M.I.T. process appear to assume optimistic ethanol yields, which have yet to be demonstrated even on a bench scale. The Penn/GE process is not sufficiently defined for a realistic cost estimate, but one was made by analogy with similar sections of the Gulf/Arkansas process (8,9).

Table VI-1

Process Cost Analysis

Process Date	Natick Oct. 1979	Berkeley 1977	Gulf/Arkansas 1981 and 1983	M.I.T. June 1979	GE/CRD June 1979
Ethanol Capacity (10^6 U.S. gal/yr.)	25	10	25	27	30
Substrate	urban waste	corn stover	urban and pulp mill waste	corn stover	poplar wood
Substrate Cost $/ton substrate $/gal EtOH	6.06 0.132	30 1.31	15.75 0.208	30.00	 0.579
Total Capital ($\$10^6$)	65.1	36.4	75.9 (1981)	34.3	37.3
Manufacturing Cost without credits ($/gal) with credits ($/gal)	 1.22 0.94	 3.11 2.73	 1.65 (1983) 0.746 (1983)	 1.05 (with profit)	 1.049
Reference	(2)	(3)	(4,5)	(6)	(7)

B. Berkeley Process Economics

Although there is not sufficient data for detailed comparison of all the various processes, a more in depth economic analysis of a single example process is still desirable to present typical process and economic assumptions and cost distributions. Because this information is available in most detail for the Berkeley process, this is the process which will be further analyzed.

The material balance flowsheet for the integrated Berkeley process (3) is shown in Figure VI-1. The process is designed for production of 10 million gallons of 95 wt% ethanol per year from corn stover. The basic process steps were described in the discussion of enzymatic hydrolysis process (Section IV,A,1). The basic design case process specifications and assumptions for acid pretreatment, enzymatic hydrolysis, cellulose production, and ethanol fermentation are presented in Table VI-2. The economic bases are given in Table VI-3.

The distributions of fixed capital cost and operating cost for glucose production are given in Table VI-4. Enzyme make-up represents the largest contribution to fixed capital (35%) and manufacturing cost (47%). It should be noted that the corn stover cost is not included. Assuming $30/ton for corn stover cost approximately doubles the final glucose cost.

Using the glucose cost in Table VI-4, the processing cost and fixed capital cost distributions for ethanol production are given in Table VI-5. The predominent portion (76%) of the finished ethanol cost is due to the glucose cost. Every 1.0¢/lb of glucose cost contributes 13.59¢/gal to the ethanol cost. The distillation cost is relatively low because a major portion of the energy for the distillation is derived from burning the residual solids from hydrolysis. Possible by-product credits for yeast cake assuming a market value of 19¢/lb (as cattle feed supplement) or 45¢/lb (as a human protein food supplement) may reduce final ethanol cost from 24.7 to 60¢/gallon. The mycellium from cellulase production,

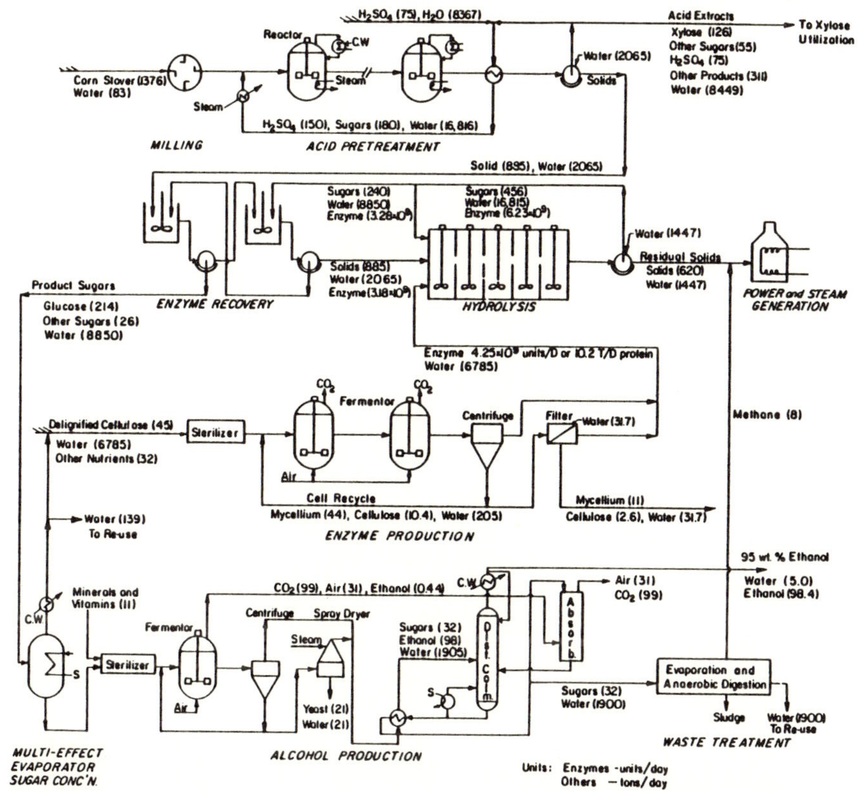

Figure VI-1. Berkeley Process. Material Balance Flow Diagram for Integrated Processing Scheme. (10^6 gal/yr Ethanol Plant).

Table VI-2

Berkeley Process Basis and Assumptions for Base Design Case

ACID PRETREATMENT:	
Feed (2mm Corn Stover)	1376 T/D
Carbohydrate Content	58%
Acid Pretreatment	0.09 M H_2SO_4, 7.5% Susp. 100°C, 5.5 Hr.
Acid Extracts (70% xylose)	181 T/D, Sugar/Acid = 2.4
Solids	885 T/D
ENZYMATIC HYDROLYSIS:	
Conditions	3.5 FPA (0.61 IU/ml), 5% susp., 45°C, 40 Hr.
Cellulose Conversion	40% to Glucose
Hydrolyzate (90% glucose)	240 T/D, Glucose/Enzyme = 21
Sugar Solution Concentration	2.6%
ENZYME PRODUCTION AND RECOVERY: (Two stage continuous enzyme production process with cell recycle):	
Inlet Cellulose Concentration	6.5 g/L
Specific Growth Rate	0.06 hr
Cell Recycle Ratio	0.8
Avg. Cell Concentration	7 g/L
Dilution Rate	0.027/hr
Enzyme Concentration	3.9 FPA (0.7 U/ml)
Enzyme Productivity	0.46 U/ml-day
Cell Yield (Mycelium/Cellulose	0.26
Enzyme Yield (Protein/Cellulose	0.24
Enzyme Recovery	58%
ETHANOL FERMENTATION (Continuous Process with Cell Recycle):	
Sugar Concentration	11.2% 90% Fermentable
Dilution Rate	0.7/hr
Cell Yield Factor, $Y_{x/s}$	0.10
Ethanol Yield Factor $Y_{p/s}$	0.46
Cell concentration in Fermentor	50 g. dry wt/L

Table VI-3

ECONOMIC BASIS

CAPITAL RELATED COST FACTORS:

(ANNUAL COST=FACTOR X FIXED CAPITAL)

Item	Cost Factor
Depreciation	0.10
Interest	0.06
Maintenance	0.06
Insurance	0.01
Plant Supplies	0.01
Taxes	0
TOTAL	0.24

LABOR RELATED COST FACTORS:

(COST = FACTOR X LABOR COST)

ITEM	Cost Factor
Direct Labor Cost	1.00
Supervision	0.15
Payroll Overhead	0.15
Laboratory	0.15
Plant Overhead	0.50
Total	1.95

BASE UTILITY RATE

	UNIT	UNIT COST	UNIT HR.
Power	KW-HR	3¢**	9875
Steam	1000 lb	32.5¢*	206
Water	1000 gal	12.8¢	196

*Self generated from residual solids
**Bought from public utility

Table VI-4

Hydrolysis Process Cost Analysis--Base Case Raw Material (Corn Stover) Cost Excluded

	MILLING	ACID PRETREATMENT	HYDROLYSIS	Enzyme RECOVERY	ENZYME MAKE-UP	TOTAL
Fixed Capital Cost X 1000 $	3,375	5,150	8,684	1,937	10,261	29,407
Annual Capital Related Costs X 1000 $	810	1,236	1,798	465	2,463	6,771
Annual Labor Related Costs X 1000 $	96	191	191	96	191	768
Annual Utilities Costs X 1000 $	109	451	657	62	809	2,088
Annual Material Costs X 1000 $	-	1,238	51	-	3,160	4,452
Annual Manufacture Cost X 1000 $	1,015	3,116	2,697	623	6,623	14,100
Glucose Cost, ¢/1b	0.72	2.21	1.91	0.44	4.69	10.0

Table VI-5

Ethanol Production Costs

Processing Cost Distribution	¢/Gal 95% EtOH	Percent of Total	Fixed $ 10^6	Capital Cost % Total
SUGAR CONCENTRATION	5.2	2.9	0.8	10.5
FERMENTATION	7.6	4.2	2.5	32.9
DISTILLATION	3.0	1.7	0.5	6.6
MEDIUM CHEMICALS	21.4	12.0	-	-
GLUCOSE	135.9	75.7	-	-
METHANE GENERATION	6.3	3.5	3.8	50.0
	179.4	100.0	7.0	100.0

Possible by-product credits:
Yeast cake 24.7¢/(60¢, if used as protein food supplement)
mycellium 13.3¢

assumed to have a market value of 19¢/lb (as cattle feed) may reduce ethanol cost by another 13.3¢/gallon.

If the corn stover cost is increased from zero to $30/ton, the ethanol cost increases 131.4¢/gal assuming that conversion efficiency is unchanged. The ethanol cost is also very sensitive to enzyme recovery. Decreasing the enzyme recovery from 58% to 40% raises the ethanol cost from 179 to 205¢/gallon.

A realistic final cost estimate assuming a $30/ton corn stover cost and allowing reasonable by-product credits is then $2.73/gallon.

C. Conclusion

As noted above many uncertainties exist with respect to the technical and economic assumptions made among the various hydrolysis and ethanol fermentation schemes. Continuing research in this area is actively under way in numerous laboratories throughout the world. Potential breakthroughs could be made at any time which might lead to major cost reductions.

A recent development which appears particularly significant is the Iotech explosive decompression pretreatment process (see Section 3, ref. #25) which promises to increase greatly the susceptibility of biomass to enzymatic conversion and lignin solubilization. Another very promising development is the Rutgers C-30 mutant strain of Trichoderma reesei (See Section I, ref. #32) which produces an enzyme system of much stronger activity toward hydrolysis of both pentosans and hexosans than previous Trichoderma strains. Further design and cost studies should be made which incorporate these advances.

The overall economics of ethanol production might also be improved significantly through effective utilization of the pentose sugar fraction and lignin. Organisms are known which will convert xylose to ethanol and other products (10, 11,12), but the development work necessary for their practical utilization has not yet been accomplished.

As a general conclusion, potential improvements appear to be evolving at a rate which should ultimately lead to economically feasible processes based on enzymatic hydrolysis for production of sugars and ethanol.

References

1. A.E. Humphrey, E.J. Nolan, "Biological Production of Liquid Fuels and Chemical Feedstocks," Report to the Office of Technology Assessment, U.S. Gov't. Printing Office, #052-003-00706.
2. L. Spano, T. Tassinari, D.D.Y. Ryu, A. Allan, and M. Mandels, "Enzymatic Hydrolysis of Cellulose to Fermentable Sugar for Production of Ethanol," Presented at the 29th Canadian Chem. Eng. Conf. at Sarnin, Ontario (Oct. 1979).
3. C.R. Wilke, R.D. Yang, A.F. Sciamanna, and R.P. Freitas, "Raw Materials Evaluation and Process Development Studies for Conversion of Biomass to Sugars and Ethanol," Proceedings of 2nd Annual Sump. on Fuels from Biomass, US-DOE, Troy, N.Y. (June 20, 1978).
4. G.H. Emert, and R. Katzen, "Chemicals from Biomass by Improved Enzyme Technology," presented at ACS/CSJ Joint Chemical Congress, Honolulu, Hawaii (April 1, 1979).
5. G. H. Emert, R. Katzen, R.E. Fredrickson, and K.F. Kaupisch, "Economic Update of the Bioconversion of Cellulose to Ethanol," presented at AIChE 72nd Annual Meeting, San Francisco, CA. (Nov. 25, 1979).
6. D.M. Jenkins, and T.S. Reddy, "Economic Evaluation of the M.I.T. Process for Manufacture of Ethanol," Report to U.S. DOE, Battelle Columbus Labs., Columbus, Ohio (June 28, 1979).
7. R. Brooks, T. Su, M. Brennan, and J. Frick, "Bioconversion of Plant Biomass to Ethanol," Proc. 3rd Annual Biomass Energy Systems Conf., U.S. DOE-SERI, Golden, Colorado, (June 5, 1979).
8. E.K. Pye, and A.E. Humphrey, "The Biological Production of Liquid Fuels from Biomass," Interim Report to U.S. DOE, Univ. of Penn. (June 1-Aug. 31, 1979).
9. E.K. Pye and A.E. Humphrey, "Production of Liquid Fuels from Cellulosic Biomass," Proc. 3rd Annual Biomass Energy Systems, Conf., U.S. DOE-SERI, Golden, CO. (June 5, 1979).

10. T.R. Batter, and C.R. Wilke, "A Study of the Fermentation of Xylose to Ethanol by *Fusarium oxysporum*, Univ. of Calif., Berkeley, LBL-6365, June 1977.

11. S.L. Rosenberg, T. R. Batter, H.W. Blanch, and C.R. Wilke, "Hemicellulose Utilization for Ethanol Production," Univ. of Calif., Berkeley, LBL-10800, April 1980. To be presented at the national meeting of AIChE, Portland, Oregon, August 1980.

12. S.L. Rosenberg, "Fermentation of Pentose Sugars to Ethanol and Other Neutral Products by Microorganisms," Univ. of Calif., Berkeley, LBL-10364, Jan. 1980. To be published in "Enzyme and Microbial Technology," 1980.